THE WORLD ACCORDING TO

KALEB

Kaleb Cooper

QUERCUS

First published in hardback in Great Britain in 2022 by Quercus Editions Ltd
Quercus Editions Ltd
Carmelite House
50 Victoria Embankment
London EC4Y 0DZ
An Hachette UK company

QUERCUS

ISBN 978 1 52942 477 5

10 9 8 7 6 5 4 3 2 1

Designed and typeset by Julyan Bayes at Us-Now Design & Art Direction
Illustrations by Julyan Bayes

Printed and bound in Italy by LEGO S.p.A.

CONTENTS

For Oscar

FOREWORD

Yeah, hi. Is this thing on? Is that how it works?

It's amazing that I've written a book. Amazing to me, I mean.
I've never done books before – reading, writing, owning. It
hasn't been my thing. I'd never written anything longer than an
invoice to a hobby farmer for rescuing whatever he was trying
to destroy that week. But I'd never done any television before
I was on *Clarkson's Farm*, either, and I didn't let that stop me.
'Give it a go' has always been my approach to anything new.
It's what made a farmer of me.

So I've done a book about everything that brought me here,
all the different bits of my life, how I see the world, and how
the world seems to see me. I still can't quite believe any of it is
happening. I'm just a young farmer, and that's all I ever wanted
to be, until I'm an old farmer. But when you decide to climb
up on the tractor, then you take the ride all the way to the end.
I hope you enjoy the ride as much as I do. Thanks for reading –
and thanks for everything!

Chapter One

Cities

I'll be the first to admit it: I don't know a lot about cities. But I do know one thing, and I'm happy to share it, because it's a very useful piece of information.

And that is: nobody should go to cities.

Ever.

Or live in them.

Or have anything to do with them.

Not my field of expertise.

Literally my field of expertise.

WHY? **WHY???**

Now, you might not think that's very practical. But I can tell you different. Before I was on the telly, I'd only ever been to London once, on a school trip. I got on the bus to go there. I stayed on the bus the whole time we were there. Then the bus went home and so did I, which to be honest was the best thing for everyone. I didn't have to bother with London, and London didn't have to bother with me.

What I like is being in a field. That's when I feel free. No one around me, and I can see for miles. In cities, sometimes you can only see as far as the next building. It freaks me out. I was back in London a little while ago for the TV show (not because I wanted to, I just had to), and I looked up, and there was a massive great big building there with a tree on top. I mean, I'm used to pruning trees that are thirty foot. But now, all of a sudden, I've got to go a thousand foot up a building, and then another twenty-five – to prune a tree? On top of a building? And then I'm a thousand and twenty-five feet up in the air. To. Prune. One. Tree.

Must have been a rush of blood to the head.

Then I went to West London, and I saw an upside-down house. What's the whole purpose of that? Why would you want an upside-down house? Why not just build a normal house and live in it like a normal person?

On my first visit to London, we drove past the grounds of Buckingham Palace and I got really excited – because of the wire security fence strung along the top of the brick wall, I thought we must be going to London Zoo to see the animals. I was totally disappointed when I found out the truth. Back then, I assumed the wire was to keep people out. These days, I wonder if it's there to keep certain members of the Royal Family in.

'Mate, I'm just a pig in a zoo, I don't know what I've got to do with any of this.'

The only things worse than trying to get onto it are being on it, and trying to get off it.

Then there's the traffic. Driving in London? Oh my God. See, around where I live, I can drive anywhere and I can raise my hand to anyone. You know, we grew up together, we work together. So, you can put your hand up to say 'hi' or 'thank you' and they'll be as happy as ever. You do that in London and they think you're a madman, or that you're trying to start some kind of road-rage incident.

It's like, here's my car, I'm in it, I've got to get to this location ASAP and anyone who gets in my way is going to pay for it. The roundabouts are enormous, and they are terrifying. It's like some kind of high-speed revolving death match on wheels.

People in London are obsessed with their cars. If my car hit their car, that's the end of the world. But with my work truck, I couldn't care less. If I drove around London in my work truck and bashed my door against a lamppost, it wouldn't really bother me.

And this would bother me even less.

I don't understand the whole thing about black cabs either. Never caught a taxi. Never been in one. Trains – never been on one. Planes – never been on one. Why would I want to do any of that? I've got a tractor.

Another thing I find really weird is that, in the city, if you want an Uber Eats or something you can just get it there and then, to your door, just like that. I can't do that where I live. I have to go out and get it myself and be independent. If people in London are sat watching TV at home and they want a cup of coffee, they never make a cup of coffee. They just get one delivered. I stand up and then go and make one. Having stuff delivered to your door like that feels wrong. If you want to get a takeaway, just walk around the corner – OK, let's not go mad here, drive around the corner – park up, speak to a few people, buy what you want to buy, then go home. I like going to the shop, having a good chinwag about everything for twenty minutes. Otherwise you're going to forget how to walk. Or at least, how to walk to the car. Or you're going to forget how to talk. Your social skills will fall apart, and you'll have trouble forming full sentences. You could end up with a city full of people who look like Space Hoppers with mobile phones attached to their hands, just ordering stuff off the Internet and having it delivered to their door.

'Is this your order, mate?'

'Yeg. Thgs.'

I clearly don't belong in a city. To be fair, it's not that different with town people in the countryside. You can spot them a mile away. I can pick out all the other farmers – you know, the sheep farmers, the beef farmers, the chicken farmers. And I can just as easily pick out the city people. They normally arrive at a farm dressed in white trainers wearing their nice white shorts, and by the time they leave they've got a brown-on-white colour scheme going on, thanks to the mud. But even without that, they stand out just because of their body language. All crunched up a little bit, all nervous. Just like me when I go to the city, come to think of it. And they don't want to step in the wrong place. Just like me when I go to the city.

Just off to pick up some chips with curry sauce.

WHEN CITY PEOPLE COME TO THE FARM

You can even tell with city people who've moved to the country, like Jeremy. You can take the city people out of the city, but you can't take the city out of the city people. And you know they're city folk when they're talking to you, when they go, 'Look at that straw,' and it's actually hay. Or the other way around. The difference between straw and hay, that's a big thing here. So many people get that wrong. Any farmer out there, it'll frustrate them. It frustrates me, anyway.

IT'S NOT HAY, TOWNIES.
And it's not Weetabix, either.

Village life is where I'm at. Village and town life. Knowing people, knowing where they grew up, where they've lived for years. It's a very small community. And if anyone needs help, no one ever hesitates to go and help them, which is a lovely thing. You go into a local shop, they know you, they've been serving you for ten years. As I go down the street, I smile and beep at everyone, like I'm Noddy.

No, not that Noddy. The one with the silly hat.

OK, the **other** one with the silly hat.

So, it's village life for me. You can keep the city. But now with all the TV stuff, whenever they want to talk to me on the telly, they want me to do it in London or Manchester. I've managed to get away with doing it on Zoom for a while because of the COVID, but when that's all over I'm going to have to think of something else. I'll just have to get them to send the camera crew to me. I'll tell them, deliver it to where I live – that's the city way, right?

There are some downsides to village life. I just can't think of any right now. Apart from snow. When it snows, you're stuck where you are. Unless you've got a tractor to get around on. I'm lucky, I have got a tractor, so that doesn't really apply to me. When I went to London, I spent the whole time looking around for a tractor, to see what the tractor would be doing in the city, just to make me feel more at home. The nearest I got was seeing one in an advert on the side of a bus.

Actually, there is another downside to village life: the gossip. You should never worry about what anyone outside your own village thinks about you. You should worry very much about what everyone in your own village thinks about you. I should know. I do it myself. If somebody does something stupid, I'll be the first to tell everyone. But at least people care.

In the city, it's like, 'Oh, did you hear about what your neighbour did?' 'I don't even know who my neighbour is.'

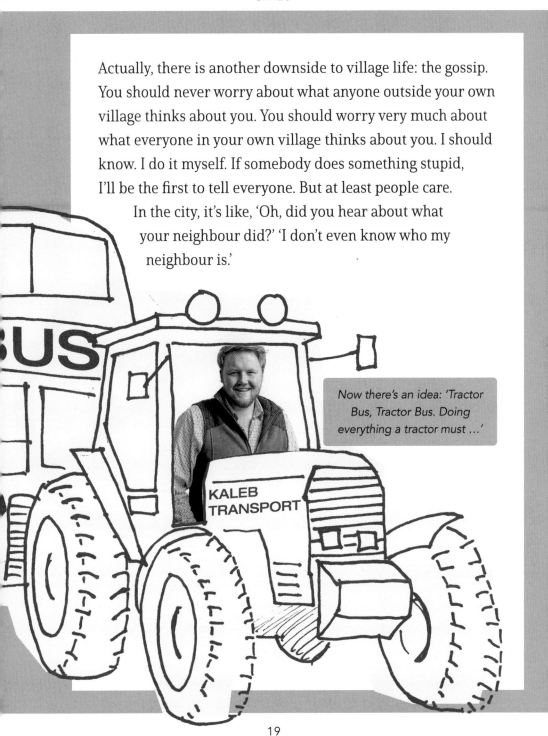

KALEB TRANSPORT

Now there's an idea: 'Tractor Bus, Tractor Bus. Doing everything a tractor must ...'

Kaleb's guide to the great cities of the world, based on zero information or experience whatsoever

NEW YORK

I don't really watch TV. I heard that some famous person once said, 'Television is for appearing on, not looking at,' and I have to agree with that. Was it Noel Gallagher? I'm glad everybody else watches it, or I wouldn't be doing this, but I don't have the time. Anyway. Apparently, New York is on TV a lot. The first thing that comes to mind is snow. Lots of snow. But if it's like London, where there are no tractors, that's no good. Can't say I fancy it, no.

Verdict: Don't go.

TOKYO

Busy. Bloody busy. You know when you send up a drone and you can see a flock of sheep? I've seen pictures of people crossing the road in Tokyo, and that's what it looks like. I'm not saying Japanese people are more like sheep than any other people, I just mean when you've got so many people all moving at once, that's what it looks like. It's like giant flocks of sheep everywhere in the street. And sheep are difficult enough here in the countryside where there's room for them. If you had a good sheepdog, you'd be all right in Tokyo.

Verdict: Don't go.
Or go, but take Lassie.

RIO DE JANEIRO

That's in Brazil, is it? The only thing I know about Brazil is football. They're very good at it. I'm a Manchester United fan, which at the moment is not like watching Brazil, because they're not very good at it.

Not at all. Even though they've got players from Brazil, but I think they must have got the wrong ones. Still, you've got to take the rough with the smooth. And I suppose there's something quite Brazilian about that.

Verdict: Don't go.

21

PARIS

The only thing I know about Paris is that my girlfriend went there for her birthday a few years ago and she lost her passport. So, I lost my girlfriend in Paris. Somebody told me that it sounded like the plot of a rom-com movie. I was actually celebrating, but I still agreed with them for the sake of appearances.

Verdict: Don't go.

MONTREAL

Is that Mexico? Canada, is it? It doesn't matter because I know exactly the same about Mexico as I do about Canada – nothing. But I know that Alaska is near Canada, and I'd love to go there. One massive area, no one around, massive kit to get around the place in, all that stuff.

Verdict: Go to Alaska instead. It's not a city, but that's the whole point.

Chapter Two

Animals

'Oh, God, not more bloody rain!'

There are basically two types of farming, unless you count whatever it is Jeremy thinks he's doing. And those two types are arable and livestock. That's plants and animals. It goes right back to the old times, to the Bible. I mean, not that I've read the Bible. There's no point; I already know how it ends. So much for 'no spoilers', eh, boys?

Giving away the ending was a Revelation too far.

But I do know that in the Book of Genesis – which is apparently named after a band Jeremy likes – you've got two brothers: Cain and Abel. Abel is livestock and Cain is arable. One day, as I understand it, they have God round for dinner, and Abel serves up a nice bit of lamb but Cain burns the carrots. God's very pleased with Abel, and Cain isn't happy about that, so he lumps his brother one. And that's that – it's never been quite right between arable and livestock ever since.

'Get off me, Cain, or I'm telling our father who art in heaven!'

26

If you'd introduced me to Cain and Abel back then, I would have known straight away which one was which. It's still the same round here. You can tell a livestock farmer because he's got a bit of baler twine as a belt around his jeans. He simply hasn't had time to go shopping. But an arable farmer has a belt with a gold buckle. And a tractor, of course. Abel should just have got on his tractor and driven away before Cain kicked off, but hindsight is 20/20 and all that.

When you're the one with the tractor, you can chase the other guy.

'Oh, come on, Artoo, if you don't hurry up, I'm going to turn you upside down and use you as a milking pail again.'

The thing about livestock farming is that it's non-stop. Absolutely relentless. Arable farming, you go all summer really hard. But a livestock farmer has to be up at four in the morning all year round because the cows won't wait. No matter if it's raining, snowing, whatever. I started off as a dairy farm apprentice. Luckily, we had two robots on the farm. For the milking. Not like in *Star Wars*.

But I also milked cows in the morning for someone else. And with animals, summer or winter, you've got to be feeding them, you've got to be bedding them. In the winter, the warmest place is in a calf. No, I know how that sounds. But when you're trying to get them to suck, you put a finger in their mouth, and when your fingers are that cold, that's luxury. Same if you're doing artificial insemination and you've got your hand up the other end of a cow. At least one part of you is warm. So you're happy to do any of those jobs. You'll jump at them.

'No, no, I'll be fine, just put me on cowshed duty for a bit.'

I don't know any bad livestock farm owners – I mean, real farm owners – because they just wouldn't survive. It takes a certain kind of person to be a livestock farmer. You've got to be hardy, and hard-working. You've got to have a massive passion for it. And you've got to like routine. I think that's why they do it. Farmers hate change. Absolutely despise it.

'I wouldn't even change my underwear if my wife didn't make me. At gunpoint.'

The robot room – really isn't as much fun as it sounds.

Let's imagine for a minute that I'm a livestock farmer. I always do every single job at the same time in the same way. Nothing ever changes. I get up at four o'clock to milk the cows. First I give them their food, then I start the equipment up, get the water, flush it all through, and then begin milking. I get to my next job at half past eight. Put my overalls and my wellies on, walk down to the parlour – or the robot room, as we used to call it – and collect the milk for the calves. When I've fed them, I bed them down. After that, I wash their buckets out. Then I give them fresh water. And then I give any of the ones that are weaning some food. Next I go and do the cubicles and scrape the yard, and then at half past ten I go back to the parlour and have a coffee.

I've been working six or seven hours already by that point. After that, I go back out and do the

Farmer or barista? Take our quiz!

31

weaned calves – like put the silage out for them – and then I start mixing the food for the cows. By the time I've done that, it's lunchtime, so I go and have lunch. Normally at one o'clock, so that'll take me to two o'clock. I've done a ten-hour day now. After lunch, I do things like dehorning calves, or tagging calves, or TB testing, etc. And then, at four o'clock sharp, I go down to the robot room again …

… get the waste milk, go back up, feed the calves, make sure they've got clean water, make sure they've got clean hay, then

start with the weaning calves again, feed them, then do the cubicle shed again and scrape out again. And I only finish when the jobs are finished. As a livestock farmer, you don't have a clocking-off time. You don't just finish at five or six like normal people in an office job. You can only finish when every job is done.

So, basically, what I'm saying is that livestock farming would be OK if it wasn't for all the animals.

The robot room – still not as much fun as it sounds.

KALEB'S GUIDE TO BRITISH FARM ANIMALS

PIGS

A pig is as solid as a rock. Pigs remind me of Dwayne 'The Rock' Johnson. They haven't got any hair on them. And they'll throw you around. Pick you up like nothing. If Dwayne 'The Rock' Johnson was on all fours ...

'Calm down!'

... that's what a pig walking towards you looks like. They are not to be messed with. They are so strong. You do not want to piss off a pig. They're lovely creatures and everything. Although I'm sure Dwayne 'The Rock' Johnson is lovely too, but imagine him walking towards you in a bad mood – it's just not something you want to see. You do not want either The Rock or a pig on all fours charging at you. If pigs get angry, that is it. Once they've made their minds up, there's no changing them.

CHICKENS

Chickens are wicked. They're honestly wicked. They've got their own little characters. We had a cockerel, right? He used to walk like Hitler. Proper marching, his legs straight out at right angles.

'Oh, adorable, am I? Don't make me come over there.'

Seriously though, chickens are pretty damn cool. They're so clever as well. I like chickens. It's probably because I like dinosaurs, and that's what they are – mini-dinosaurs. A whole herd of little T. rexes running towards you. Really! They're descended from dinosaurs, or at least pterodactyls, or something like that. So they're like little dinosaurs with feathers.

'Nein! Ist nicht "the goose step" – ist "the chicken step".'

I think it went T. rex → dodo → chicken. The T. rex would be wondering how that happened. If a chicken had little hands, like a T. rex, just imagine how cool that would look.

Are you not entertained?

And they're brutal, as well. They're proper brutal. They have the T. rex in them, still. I know it. I saw one the other day kill a frog and a mouse. The trouble with farming them, though, is predators. I spend my life watching out for foxes: getting the chickens back in, getting them locked up, unlocking them, feeding them, watering them, all the while looking left to right for a fox.

I started my whole business with chickens, so I love them. Without chickens, I wouldn't be doing what I'm doing now. Which is still farming chickens, I grant you. But farming them makes you very content. It's peaceful, farming chickens. You spend your entire life counting. If you've got thirty chickens, you're not going to make a lot of money. So you have to manage about three million, and you need to count them in and out all the time. And then you spend your life adjusting the temperature of the shed because of the humidity. Turning dials and counting. It's a long way from *Jurassic Park*. Thank f***.

Don't move. She can't see us if we don't move.

TURKEYS

The one thing I know about turkeys is that you've got to have two people whenever you go in a shed full of turkeys. You can never go in on your own because they can trample you and kill you. So next Christmas, when you're tucking into your turkey, just remember – they could have got you first. Some farmer risked their life getting that turkey for you.

Fowl play is suspected.

SHEEP

I don't have sheep. I won't have sheep. I don't want sheep. There's a reason for that. I don't have sheep because sheep are F***ING AWFUL.

Just in case I haven't quite been clear on my views here, let me add for the record: I. F***ING. HATE. SHEEP.

I take my hat off to anyone who farms sheep. You give a sheep a choice between life and death, and it'll look at one, and then at the other, and then it'll run towards death every time. Every day, their main objective is to die.

'Well, we're not crazy about you either, mate.'

'Deliver me, Satan, from this mortal plane!' *insane guitar shredding intensifies*

They have no other objective in life. They wake up in the morning and they go, how am I going to die today? I know – I'm going to shove my head in that gate, turn around four times and strangle myself. The grass is always greener on the other side to them. So they'll happily kill themselves to get there. I really f***ing hate sheep. I. HATE. THEM.

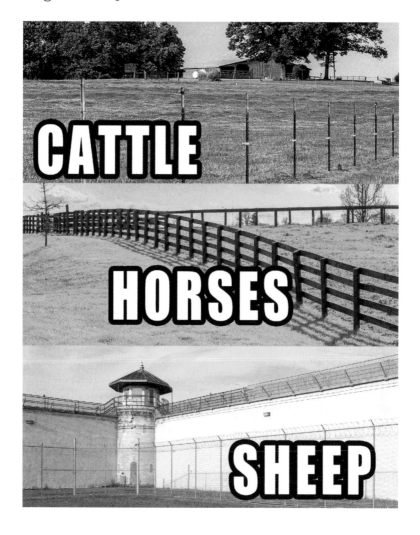

COWS

I love cows. In a herd, there's always one who loves you, who wants to come up for a bit of a scratch. There's always the one who stands out. There's always the one who's funny. There's always the one who makes you laugh all the time – you look at her and she does something weird and cracks you up.

' ... and the ones who get together and form a solemn delegation.'

There's always the one that's really placid. And then ... you get the one that's a bitch. The one that looks at you and goes: 'I f***ing hate you, I literally f***ing hate you.' And you're like, 'I've done nothing wrong!' You just know, when she looks round at you and gives you those eyes, she's thinking, 'I'm going to make your life a living hell. But you're still going to love me for it.' And you weirdly do. Because without that personality, it wouldn't be the same.

The phrase 'nosy cow' is literally true. They're so nosy. They're the nosiest things ever. They just want to see what everything is. They're not really scared of anything. If, for example, you're doing a bit of work in their shed or something like that, they always have to come over. I've been on a ladder in their shed trying

That nosy cow from next door is back.

to fix a light, and they've come across and knocked the ladder over, leaving me hanging up there. I suppose at least the sheep only try to kill themselves, and not you. I still love cows, though.

TROUT

Trout are just sheep with fins. They're just as stupid. I know a couple of local farmers with trout. Trout are just trying to destroy themselves, only underwater. And you can't do anything about it. At least with a sheep you can see when it's happening. But with a trout, the first you know about it is – it's succeeded in destroying itself. You think, how have you got yourself in the air, five feet up, trapped in a net?

'Hooray! Death awaits me, somehow or other.'

DOGS

Right this second, I'm looking at what you'd describe as a typical collie dog. She's not working with sheep or cows or anything like that. Right now, she's sat with a ball, in my barn, staring at it. Continuously staring at it. I've got a Labrador puppy right here as well, and he's just forever running around, picking something up, carrying it around, dropping it, then going back and picking it up again. And the collie is still focused on the ball. Now the Labrador's looking at the collie staring at the ball. God help him if he tries to pick it up, though.

'You can touch the ball. Or you can live. Your call.'

They're hyper-vigilant, collies. Bred that way. Concentrate on things. Watching a collie work is a bit like watching a gold bar running around a field in circles. You have to make sure no one nicks it, because everyone knows how valuable they are. There was a sheepdog that sold a while back for £20,000. Worth every penny.

The dog's job is to watch the sheep, and your job is to watch the dog. It's like a production chain. You follow it down the line. Normally, the farmer's at the bottom and there's the middleman and there's the person who sells the stuff at the end. This time, the farmer's actually the middleman. He's basically hiring the dog to do all the sh*tty work.

'We've assembled here today to announce that we're forming a trade union.'

GOATS

I can't really see the point of goats. Goats are just … goats. A goat's purpose in life is to have some horrible-tasting meat on it to go in a curry. Curry was designed for that: to make goats' meat taste just about edible in the desert. They're

'Oh yeah? Bite me.'

a proper good clear-up animal, they eat everything – no wonder they taste horrible. I mean, I like a bit of goat, to be honest. But you've got to mix something with it. You can't just have a goat steak, can you?

GEESE

I haven't got much experience of geese. Thank God. They're f***ing brutal. They're like guard dogs. You have geese on a farm to intimidate intruders. Intruders aren't scared of the dogs. The dogs are all nice and everything these days. But if you've got geese there, the intruders know to stay in the car and not move a muscle.

'Finish him!'

Chapter Three

Heroes

'Got me arm caught in a thresher.'

Heroes come in all shapes and sizes. But for me, the ones who stand out the most are the ones who are determined. The majority of farmers are heroes. They always want to help other people. OK, sometimes they want to help other people get off their land.

But if you're part of the farming community and you need help, no matter the day, the week, the time, they'll do anything to help you. When I was eleven or twelve, my mum and dad split up, and it was a rough patch in my life. Farming saved me.

I started farming when I was thirteen. Without it, I think I would have been just a typical teenage guy who wants a normal nine-to-five job, goes to the pub every Friday and Saturday, and then does it all again the next week. Most people who start their own business don't do it until their late twenties. I started one when I was sixteen.

'So did I, but I had a hard paper round.'

My biggest heroes are the people who have helped me along the way. David Haine was the first farmer to take me on. That's how I started out, on the weekends, as a dairy farmer. Without him saying, 'Yeah, come and work for me', I wouldn't be where I am today. Which is on top of a tractor. And that's exactly where I want to be.

'I'm the king of the world!'

Then there was Howard Pauling. He gave me confidence, and he trusted me with responsibility, which was a big step forward for me. He was so passionate about farming. We would start jobs at eight o'clock at night. I think that's where I got my work ethic from. And David Haine was so caring and loving towards me. When I had just started my own business, my tractor broke down, and he helped me fix it, and never charged me. That would have broken me, right at the start, if I'd had to pay for it.

There isn't a tractor joke here. Or any joke. Because that's a nice story and we just want to take a moment.

What's really sad is that the only two farmers I ever worked for before starting out on my own – both of their farms got sold. It made me feel like I must be bad luck. I felt like a jinx. A Jonah. I probably beat myself up a little bit too much.

I still do some work at one of those farms. A millionaire bought it, and when I go into the yard it still upsets me. I remember all the fun I had, all the work I did, how kind the former owner was to me.

So that is why anyone who is a farmer is a real hero to me. Somebody who was there for me and really helped me on the way, gave me the biggest uplift in life. People might think that's soppy but it's the truth.

Just a minute. We've got something in our eye.

So now I've been chucked into this crazy world of television. And without people to advise me, I wouldn't be where I am today either. Which is on top of a tractor, on TV.

'I'm still king of the world, now streaming on a certain media platform near you!'

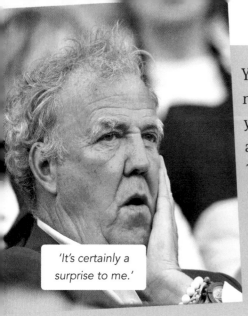

'It's certainly a surprise to me.'

You know, there's a lot to TV. It's not just turn on the camera and you're there. There's a lot to think about: the whole directing business. There's so much pre-planning and thinking ahead. 'Have you got this little cutaway? Have you got that?' We've done a whole series, but I'm still learning. Mainly, that means learning from Jeremy. I mean, he's a complete f***ing pain in the butt when it comes to farming. But when it comes to television it's like the roles are reversed: he's the one who knows what he's doing, and I don't. People might be surprised by this, but when it comes to television, he's a hero to me.

So, when I think about heroes, those are the people who come to mind first: David, Howard and Jeremy. But most people have heroes they don't actually know in real life, and the only one I can think of is Jordan Belfort, the guy whose story was in *The Wolf of Wall Street*. That's my favourite film. I found it inspiring, the way he was so determined to make money and be a success. A lot of people have told me I'm taking the wrong lesson from that film,

'… and now we move on to seeding barley.'

though. I mean, maybe all the crimes, drugs, violence, car crashes and going to prison aren't ideal in a role model. But you've got to admire the guy's drive and ambition.

'To be fair, I've never heard of you, either.'

I've had people ask me what I think about Greta Thunberg and whether, as a young person, I find her inspiring. But the only thing I know about Greta Thunberg is that, apart from people asking me what I think about Greta Thunberg, I've never heard of Greta Thunberg.

And sometimes people talk me to about old heroes of farming, but I don't know anything about them either. 'Oh, do you know so-and-so, he invented the Wanking Lizzie', and things like that. And I'm sure it was all very useful and important back in the Stone Age or whenever, but this is the twenty-first century now, and the only opinion I have of Jethro Tull is that I've never been a big fan of that whole hopping around playing the flute business, myself.

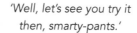

'Well, let's see you try it then, smarty-pants.'

I heard that Jeremy once presented a TV show about a hero of his, so I looked it up and it turned out it was part of a whole series: *100 Greatest Britons.* I don't even know a hundred Britons, let alone great ones. I've just dipped into the list to see if there were any that rang any bells.

KALEB'S GUIDE TO GREAT BRITISH HEROES

SIR WINSTON CHURCHILL

I think he was the president. Or prime minister. And what I remember is that he told all the farmers to do rewilding – you know, let wild meadows grow. And then the Second World War happened and we didn't have enough food to feed our armies. We were all on rations. But we would have been fine if we hadn't started growing all this grassland crap. Or maybe that wasn't him. But whoever it was, they were a twat and they shouldn't have got a knighthood.

Right back atcha, pal!

ISAMBARD KINGDOM BRUNEL

This is the guy Jeremy was talking about in the TV show. Apparently, he was a one-man industrial revolution. Bridges, railways, ships. Although I found out he didn't do any tractors, so it's a thumbs-down from me. He's got a weird name. Imagine having a name that long – having to spell that out every time. 'We've got this bridge, designed by Isamdiddly Rumpotato Biddley-Boom-Sasquatch.' It would take him longer to write down his name than it would to build the bridge.

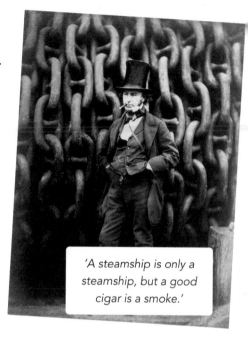

'A steamship is only a steamship, but a good cigar is a smoke.'

CHARLES DARWIN

I thought he was a book writer. And I was right about that. And I thought I'd read a couple of his books – children's books about farms. But it turned out that I'd got that bit wrong. I was doing really well up to that point. But it's actually even better: he's the theory of evolution guy – all about dinosaurs and stuff like that. So right away I wanted to meet him. Unfortunately, I found out he was dead. Then I saw he'd written books called things like *The Descent of Man* and *Selection in Relation to Sex*,

which sounds a bit frisky. I assumed he must have been a bit of a boy and maybe died in some horrible drug accident. But no, he lived to a decent age and died of natural causes. Must be because his brain was tuned in to dinosaurs. I reckon we should bring him back. After all, if you can do it with dinosaurs in a bloody big park on an island, why can't we do it with Darwin?

WILLIAM SHAKESPEARE

Oh, f***ing hell. I can't stand that name. In school, every time I heard 'Shakespeare', I just got bored. I switched off. Stratford's not far from me, so I'm always hearing about it. I once asked somebody, was it Shakespeare who invented English? And they looked a bit surprised and said, yeah, it sort of was. I remember the Three Witches. Awful. Every time I got into English class, it was the Three Witches. We even had to watch videos of them casting spells and saying weird sh*t, and I'm thinking, 'Oh, for f***'s sake, turn this crap off. Just teach me how to spell "happy" – or "tractor", which amounts to the same thing – and tell me where to put the full stops and commas. Don't tell me how the f***ing Three Witches mix up their spells. Who cares about that?'

'Sheesh, girls, tough crowd.'

SIR ISAAC NEWTON

With this one, I thought, 'The name rings a bell. Did he have a hump and live in a tower? No, that's Audi Quattro.' Then I thought, 'Was he a scientist?' and yes, that was right. He discovered gravity. Imagine discovering gravity! 'Go on, float, or I'm going to call it gravity.' It's like chickens. Imagine the first person who discovered a chicken egg. Who the hell thought, 'You know what, that's just come out of its arse, I'm going to eat that.' Same with cow's milk: 'I'm going to drink that stuff that comes out of that udder.' These are the people who move civilization forward, though. So he's all right by me.

Not only that, but Newton discovered which one came first.

ELIZABETH I

I thought, 'Is that the queen now, the one in the palace in London?', but the queen now is Elizabeth II. Which got me wondering, why do they have the same name? Why don't they change it up a little bit? With cows, in a dairy herd, you take the first letter of the mother's name and use that to name the calves. I think that's a good way to do it. So instead of getting two Queen Elizabeths, you would get one Queen Elizabeth and one Queen Edith. Or maybe a Queen Ermintrude, if you really want to go with the cow thing, but I don't want to be disrespectful to the Royal Family. I'm just saying a bit of variety never hurt anybody.

'Kneel before the glory of my majesty, dear hearts.'

JOHN LENNON

I guessed: footballer. I guessed wrong. I've heard of the Beatles. I'm more of a Who fan, though. I love the Who. 'Out here in the fields ...' My grandad, who sadly passed away last year, got me into them. He was a massive Who fan, so I was a massive Who fan too. Every time we got in the car, we'd be singing away.

'Behind Blue Eyes' and all that. What I discovered over time is that, just as football teams have fans who hate other teams – like if you're Man United you hate Man City – so if you're a Who fan, you hate the Beatles. Now times have changed, and everyone loves every singer. But there was rivalry back then. Like Michael Jackson and Queen. You either liked one or the other. It's like how you either like cows or sheep. Cows are definitely the Who, and sheep are definitely the Beatles.

'Baa, baa, baa, baa baa baa baaaaa, baa baa baa baaaaa, Hey Jude!'

'Mooooo are you, moo moo, moo moo?'

HORATIO NELSON

He was an important sea captain. And he beat a guy called Napoleon, who was French and basically declared war on all of Europe. Which is brave. Stupid, but brave. I've never seen the sea. But I have been to Trafalgar Square, where there's a statue of Nelson on a big fence post. I wouldn't want to be the one who had to put that up. I drove through Trafalgar Square. It was

'I have nothing up my sleeve.'

the worst day of my life. To be honest, I was concentrating more on not crashing than on looking at what was on top of that post. I was more scared that I was going to get run off the road and crash into him. Try explaining that one, knocking over Nelson's Fence Post.

OLIVER CROMWELL

I remembered something about a guy called Charles getting his head chopped off, and I found out it was because this other guy, Cromwell, started an English Civil War with him. It doesn't sound very civil to me, but I don't write the history books. I definitely know the type: a stupid twat who wants to start fights in England. It's not as if you don't get one of those everywhere you go around here. I'll bet you this Cromwell character was a regular in a pub called the Red Lion. They always are.

Oi, Cromwell! You're barred!

DAVID BECKHAM

I've got to be careful what I say. He lives almost next door to me. He's another Londoner who's come out of the city to live in the countryside. He lives down this way, right next to a place called Soho Farmhouse, funnily enough. I wonder why. He's got a massive house, and ponces around dressed like a farmer. Or at least what he thinks a farmer looks like. Stick to what you're good at, mate. Take the checked shirt off and put a plain shirt back on.

Not sure the playing conditions are what you're used to, mind.

ALEXANDER GRAHAM BELL

He's the guy who invented the telephone. And he probably shouldn't have. My life would be so much easier if he hadn't gone and done that. The number of phone calls I get is unreal. He's ruined things for children, too. I bet he didn't think of that beforehand, what it would do for kids. Go back and uninvent it – that's my advice.

'All right, Kaleb, mate? Just calling to see if you fancy a pint later.'

DONALD CAMPBELL

Now we're talking!

A daredevil and a high-speed record-holder, so I read, driving cars and boats. I can't say I was very impressed. Now, if he'd done it on a tractor … that would be more interesting. Plough a field in record time. Anyone who did that would be a hero to me, definitely.

ALEISTER CROWLEY

He was devoted to devil worship. Christ almighty, the world didn't need him, did it?

GUY FAWKES

Thanks to him, we have Bonfire Night. It's an amazing night – means I can get rid of all the stuff I don't want any more. It just opens up this massive opportunity. Everyone's doing it, having a bonfire and throwing it all in, so

What have you come as?

everyone looks away. Every farmer's dream, Bonfire Night. They love it. So, he didn't die in vain. He went for a good reason. Thanks, Guy!

Never you mind what's in it. All you need to know is, it's not coming back.

SIR DAVID ATTENBOROUGH

I love him. I love his voice. And the knowledge he has of animals – it's amazing. He should be king. He should be up there on a throne. I can just see myself meeting him and bowing to him. I used to watch his programmes all the time, because I wanted to learn about animals. That's what I spent my childhood watching on TV. Never any kids' programmes, always David Attenborough. Him looking at fossils and explaining them. He was a proper hero and still is. I want to teach him all about farming, though.

Sir David deals with worse things than sheep before breakfast, while trying to avoid becoming breakfast.

Chapter Four

Hairstyles

Uncanny, isn't it?

I have a philosophy about my hair. I get a lot of stick because I keep changing my hairstyle. But there's a good reason why I do. My dad, my gramps – all the men in my family – have not got a lot of hair left. As I've already said, us farmers hate change. And I'm going to keep on saying exactly the same thing because … well, see if you can guess.

But the one thing I like changing is my hair. Because I know I'm going to turn out like my grandpa and my dad in the long run. So I may as well be seen with lots of different hairstyles. I'll try them all until I find one I *really* like. Then I'll stick with that until the inevitable happens. I just hope I find it in time. I might have it for five minutes and then get an egg in the nest, and it'll all be over.

Don't knock it, it's a look.

I thought I might have found The One with the perm. I like the perm. In fact, I love the perm. But the problem with the perm is, it was killing my hair, because it's not natural. I wonder how many pop stars and footballers from the seventies and eighties lost their hair because of it.

'I regret nothing!'

So that's why I had to change it again. I'm going to go with the noughties mullet next, like Cristiano Ronaldo had. A little bit curly and slightly long at the back. I'm not going to attempt to do the long mullet because Gerald owns that haircut. His is amazing. It would really knock my confidence to be shown up by him. So I'm not even going to try.

If you can't compete with this…

… then go for this.

This in the picture on the right was the haircut I had back in college days. I thought it was really cool. I couldn't grow a beard, though. I used to have my hair long in the middle and cut short on the side. Then I used to sweep it all over with some sh*t greasy hair gel. At the time I thought it was wicked. And looking back

at it, I was right – it is wicked. I was doing Level 2 Agriculture, but I'd say that was easily a Level 4 haircut.

Then this happened.

Wait, isn't that a mangel-wurzel?

Here, I was growing my hair out to get ready for the perm – it had to be a certain length to perm it. We were at our harvest, and I had an Alice band on at the time, and I thought, 'Bugger this, I may as well gather it together into a top knot, like Gareth Bale.' He was my style inspiration. There was only one problem: it was awful. I hated that hairstyle. It was probably my worst one. It didn't make me any better at football, either. Although I'll admit it was more aerodynamic for riding my tractor.

On the right: this was a messy look. Not a Messi look. I do imitate a lot of footballers' haircuts, but not that one. Anyway, I can't do fades, because I'm blond. Or ginger. Whichever way

Are you lo-o-onesome tonight ...?

you want to look at it. I like to think I'm more blond. Strawberry blond, at a push. If I do a fade, it just makes my hair look the same as my skin because I'm so pale. It's like I've already given up on the whole idea of hair at all. For this haircut, the way it worked was: get up in the morning, put a bit of 'hair dust' in there, fluff it around a bit, then leave it, and that's your style. I think I put that filter around the photo to try to rescue it, but let's face it – it was already far beyond that.

Well, hello there.

I call this the Summer Kaleb. The summer months really bring out the blond in my hair, like I've dyed it, or even bleached it. WHICH I DEFINITELY HAVEN'T. It's a haircut that's layered and all pushed to one side. It's embracing the natural highlights. This chapter is turning into a cracking beauty guide, now I come to think of it. Gentlemen readers, please be sure to pass this book on to the ladies in your life, and thank me later.

Below: this is what I'm going to look like one day. And you know what? I'm going to embrace it when I get there. I'm not going to go for those stick-on things – I think I'd be seen as a bit of an idiot. Some stubble may stop me looking quite so much like a boiled egg, mind. Not a full beard, though, because then you just get all the jokes about your head being on upside-down. Although a good beard does make the people who can't grow one jealous.

I have seen the future, and it hurts.

I keep my flat cap licence with my driving licence, behind the sun visor.

There's always hats – as long as you've got a hat face. I can't wear baseball caps, but I can wear a farmer's hat. In fact, only farmers should wear farmers' hats.

If you're not a farmer, or at least a country person, and you wear a hat like that, you're a total twat. I'm looking at you, *Peaky Blinders* fans.

Here I am below with my other half. I had the perm, and I straightened it to a longer perm length. It has a bit of a wavy look. But it's weird to have one big curl in all the straight bits.

The actual Peaky Blinders, though – they can wear what they like and they won't hear a peep out of us. Sorry.

Later, I had tramlines put in either side of the perm. I was going to a farming event and I wanted to look more like a farmer. But apparently no other farmers have them, so that didn't go down too well. You wait, though. I'm just too fashion-forward when it comes to farming style. They'll all have them in a year or two.

Aww.

Did we already say 'Aww'? Because if not, let us just add, 'Aww'.

This was my lockdown hair. I worked out I could push it over to one side. It all went a bit emo. So this is basically the Lockdown Emo Farmer look. I never went any further with this because I was worried I might start listening to My Chemical Romance and drinking cheap cider in graveyards. Although you can see from the next picture I was already getting a bit sulky.

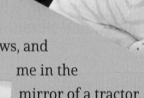

'Can't wait until the pandemic is over and we can go and see the Lockdown Emo Farmers on tour.'

Now I think of it, you never see an emo farmer. But the good thing is, no matter what you look like, a cow's not going to take any notice. That's one of the reasons I like having all these haircuts. Before I was on TV, nobody ever saw them except for a hundred cows, and me in the mirror of a tractor. There's no such programme as *Steer Eye for the Straight Guy*. A cow's not going to judge you.

'That's what you think, pal.'

My hair went a bit Afro after lockdown, when I couldn't get to the hairdresser's and I was trying to control it. Not that you'd know it here. The perm trains your hair to grow in a certain way, even after you've cut the perm off. This was me trying to control that before it happened. It just sticks up like a cockerel's arse. I should have let the farming Afro just happen. There ought to be a word for it. I'd call it a Jethro.

'I haven't slept in seventy hours, and you think I should be worried about my hair?'

I wanted to prepare like a proper footballer for this charity match

My perm, my son, and Oxford United – I'm living the dream!

at Oxford United, so I went to the hairdresser. Never mind training, get your hair done, like a pro. Or like a cow – before you get milked, go to the brush in the yard and have a bit of a scratch. Cows are a very big influence on me. This was as good as my perm got. A perm's got to fit the shape of your head. This, on the left, was a geometric perm – almost square. This is peak perm.

I still have a lot of haircuts to go, though, before I find the right one. I'm like a prince with a glass slipper looking for my Cinderella, if the glass slipper was my head and Cinderella was a hairstyle. I've even got my hairdresser to do me a list of classic haircuts to look through and see if I fancy any of them one day.

KALEB'S HAIRDRESSER'S GUIDE TO GREAT HAIRCUTS

THE PRINCESS DI FEATHER CUT

This looks like it would take a lot of controlling and a lot of styling. I haven't got time for that. I'm a farmer. I'll admit that for a farmer I do spend a lot of time controlling and styling my hair, or else I wouldn't be writing about it now. But not that much. OK, not quite that much.

THE ANDY WARHOL PLATINUM WIG

I've more or less tried this already. It's near enough what I had with the perm. I like the separation, covering the bald bit on the side. And on the top. And on the other side. And, I'm guessing, the back, although I can't see that – maybe he went for some kind of wiggy undercut. I do see him as a kind of prophet, though In the future, everybody will change their hairstyle every fifteen minutes.

THE MOHAWK

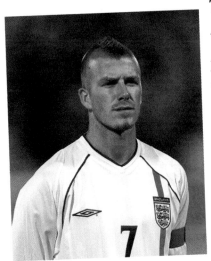

This style makes you look like a pig that's been rubbing up against trees too much. You lose all the hair on the side but keep the bit down the middle. I like to take inspiration from cows – but not pigs. Not even if David Beckham thinks it's a good idea. He probably hasn't met a lot of pigs. If he had, he'd have thought twice about this one.

THE MOP TOP

This makes me want to put a counterweight on it, to make sure it's balanced out. Whenever you put a heavy attachment on the back of a tractor, you need to put a front weight on, to stop the tractor flipping up. You'd need to do the same with this haircut, only with a back weight. Otherwise it'll just flip straight over and cover your mug like some kind of hairy alien facehugger.

THE CAESAR CUT

I like this. It's wavy. In fact, I've pretty much had this haircut already. I've crossed that Rubicon. The die is cast.

THE BEEHIVE

Not very appealing to bees, to be honest. If I was a bee, I wouldn't want to go in there. Which is probably for the best, being in the country. You don't want to be chased all over the fields by swarms of bees. Either way, I'm not about to risk it.

THE RACHEL

Ah, she's so fit, isn't she? Sorry. What was I supposed to be writing about? I like that she has become a hairstyle. I want to be like that. I want people to go into the hairdresser and say, 'Give me a Kaleb cut', whatever that turns out to be. Obviously, I haven't decided yet, which is what all of this is about. God, she's fit, though … where was I? Never mind.

DREADLOCKS

This is definitely not my thing. Dreadlocks remind me of the Island Boys, I've watched online how you do it and I can see hairstyle maintenance issues with this one. I have to take a functional farming approach to these things.

THE PRINCESS LEIA

Again, that's too much hard work. I don't have that much hair, and I don't have that much time. In fact, I'll never have that much hair and I'll never have that much time. I'm only going to have less of both as life goes on.

THE MARCEL WAVE

I really like this one. I wouldn't mind that at all. That would be smooth. I reckon I'd rock that one if I could do it. That's inspiration right there. Kaleb Cooper: matinée idol.

THE POMPADOUR

This is a definite no. The volume of that would not be good for me, getting out of a tractor. My hair would be touching the top of the cab. I've got to be practical about these things, and that's definitely not practical.

THE COMB-OVER

I'm going to have to make a decision about that one before too long, aren't I? The trouble with the comb-over is – one gust of wind and it becomes the flyover. Off it goes. Imagine me on a quad bike with one of those. Again, it fails the practicality test.

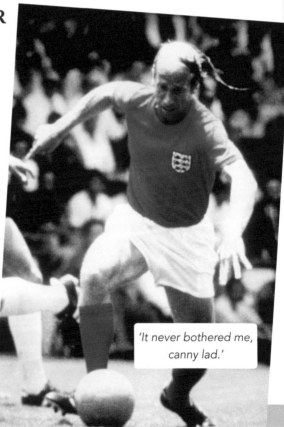

'It never bothered me, canny lad.'

THE GEISHA

This one's smart. Again, I haven't got that much hair, and I never will, but I do like it. I can't see myself doing the ornaments, though. They'd just get snagged in everything and the cows would want to eat them. Also, I'm not big on tea ceremonies. Tea, yes – ceremonies, no. I haven't got four hours for a cup of tea. I'm lucky if I get four minutes.

THE HIPPIE

I think there's probably days when I look like this without trying. I had long hair when I was a kid, down to my shoulders.

I had the proper flick and everything. I wouldn't go back to it, because I was too scared to chop it off. And then when I did it was amazing. I could actually lift my neck. I never looked back. Although I could look up and around, so that was progress.

THE FLOCK OF SEAGULLS

What the f***'s going on there? Imagine me ploughing in that!

THE DAY-GLO BARNET YOUNGSTER

I wouldn't dye my hair a bright colour. Or any colour, come to that. I very much identify as blond. It's an ongoing fight. People say, 'You're ginger.' And I reply, 'No, I'm blond.' They keep mis-hairing me, but I won't have it. I'm here. I'm fair. Get used to it.

THE YOUNG JEREMY

All right, I can't deny it: I'm envious. That's a natural perm. If I had a natural perm, I'd be all right. We all wish we had hair like that, don't we? If not the face.

Chapter Five

Technology

'Mwahahahahahaha! Just wait until Farmers Weekly *hear* about *this*!'

Technology scares some people. But it's only a problem when it's in the wrong hands. Specifically, Jeremy's. I love technology. And I love machinery. Let's be honest, as you know full well by now, what I really love is tractors. Or anything that looks even a little bit like a tractor. Or reminds me of a tractor. Or is in the slightest way associated with a tractor.

I LIKE MY
TRACTOR
AND MAYBE
THREE PEOPLE

Tractor bumper stickers: definitely a thing.

When I'm driving a tractor through town and I see a mum with a kid, the mum will always say, 'Look – tractor!' And the kid stops everything they're doing and stares at the wheels going round with an expression of complete amazement. Well, I'm still like that. At twenty-three years old, I'm basically that kid who's been allowed to have his own tractor.

Meaning of life: unlocked.

I love *all* tractors. Even the first one I bought when I was fifteen, with £5,500 I'd saved up by working my arse off. I didn't bother to test-drive it because I was so excited, and it broke down almost immediately. Of course, I wasn't allowed to drive it on the road until I was sixteen. So I did not do that. Absolutely not. That would be wrong. And anybody who says I did is either lying or has their dates mixed up.

I can't believe how quiet the roads are this morning. I hardly see any traffic at all.

And they never saw anything like this, either.

But you're allowed to drive a tractor on the road a year before you're allowed to drive a car. At sixteen, I already had a little car my dad had bought me. And I worked out I could drive a tractor and a low-loader on the road, get my car up onto it, drive it to a farm, then take the car off and drive it around there. Tractors. Truly, there is nothing they cannot do.

'This is the single greatest day of my life! Apart from all the other days when I also get to ride a tractor.' I can totally relate to what this boy is thinking.

The way we measure tractor power is in horsepower. So imagine it like this: let's say you've got a 300-horsepower machine – that's literally 300 horses' worth. Think of all of that in one machine. That's what I see when I look at a tractor.

So beautiful. So wild. So free. So tractor-y.

If you've got a Lamborghini – an actual Lamborghini, not a Lamborghini tractor – or another posh sports car, and it's worth, let's say, £180,000, and you pull up in front of a nightclub, you'll look rich and flash and all the girls will love you. Now, my tractor is worth more than £200,000, but it doesn't have quite the same effect. Clearly this is an attitude that needs to change, and if I have my way, it will.

'*sigh* ' This is all very well, but how am I supposed to compete with that Kaleb fella?'

Plus, I can park a tractor wherever I like, because who's going to stop me? And ladies, a tractor can fulfil your deepest fantasies. It's even more romantic than riding away with the hero on a horse. It's like riding away with the hero on 300 horses. Only in an air-conditioned cab.

Also, having ridden just one horse, I can testify that no bloke is in a position to do anything romantic afterwards. Another big advantage to the tractor. Trust me, its time will come.

'We thought we'd go back to where it all began.'

All the same, I have to admit there's more to life than tractors. There's gear that you can hook up to tractors. And there's gear that makes you wish you were on a tractor instead.

Like the mini-digger. I have a love–hate relationship with that. Sometimes I'm really good on it, and sometimes I'm awful. I like digger work but then I lose concentration. I'm a farmer and everyone expects me to do it, which is fine, but I'd much rather be on a tractor going up and down a field than be in your garden digging a septic tank. I was thinking of getting that made into a bumper sticker for the mini-digger – 'I'd Rather Be On My Tractor' – but then I decided it wouldn't look very professional.

In fairness, if you can't have fun with this thing, do you really know how to have fun at all?

The truth about technology is that it's OK for work, but when it comes to play, it's awesome. Take the mini-digger. You can lay a plastic sheet down, squirt some Fairy Liquid and water on it to make it nice and slippery, tie a rope to a beam, and start doing doughnuts with it, while holding onto the rope. But you didn't hear about that from me.

I've got three toppers, and that's what started me out. I bought a flail topper at the age of fifteen along with my first tractor. I do lots and lots of topping. I love topping. It's so satisfying.

When you're taking out nettles or thistles or long grass, it's one of the most enjoyable things you'll ever do in your life. You aim for one thing and drive over it just because you can.

'Yes, I too love to do lots and lots of topping. It *is* satisfying.'

Don't get me started on potato harvesters,

Oh, right. Yes, that looks waaay more enjoyable.

though. I've only ever done potato harvesting on the TV show. We use a rust-bucket from around 1930. The very worst thing about it is: it still works. It's agony to operate it, but because we're only doing a small acreage to supply the farm shop, we don't have an excuse to buy a comfortable new one. I like the old kit, but when I'm doing nine hours a day on it, I stop being so much of a fan. It gives me a proper taste of old-school farming. Farmers must have been a lot tougher back then. I take my hat off to them. It's worse than riding a horse. I think I should get compensation for the possible damage to my fertility.

Coincidentally, also what the contents of your underpants feel like afterwards.

I've also bought a mole plough that lays water pipe without you digging it in. I can do a kilometre and a half a day. It leaves one incision in the ground. Then you just fold it back over with a tractor wheel, and you're done. It's an example of the simplest machine that does the most amazing job. That's the kind of technology I like the best. You can always tell a hobby farmer because they go for the most complicated machine that doesn't do anything useful at all. Not that I'm mentioning any names.

Pictured: inventor of the mole plough.

Perhaps a bit too much of a hurry.

On a completely unrelated matter, Jeremy's got a Supacat. It's an army vehicle. Getting in and out of it is virtually impossible. I don't know how they do it, but if I was in the army, I'd want to be able to get out of whatever I was riding in very, very quickly. On the farm, it's basically a toy. Or an ornament.

I'll stick with my quad bike. Which, I can't deny, I also use as a toy when I'm not using it for work – I use it for spraying, for instance. But at least I can get off it in a hurry.

KALEB'S GUIDE TO THE HISTORY OF FARM MACHINERY

JETHRO TULL'S SEED DRILL (1701)

Now that is something I wouldn't want to use. Hard to turn. Big ugly wheels. It was better than just scattering the seed by hand, I suppose. But if you're planting a fifty-acre field, imagine doing it with that thing. It's going to take you a little bit longer than nowadays. Back then, they had more people working in a field, and you can see why – they needed them! But this seed drill was the best machine they had.

It amazes me that we can go from people pushing something like that to a machine that slots the seeds on its own, with an exact number per square metre, each one planted precisely to the millimetre, travelling at fifteen kilometres per hour. At one end it's just some cogs turning slowly, to match your speed, and at the other, a computer. That's what happens when you match them up: bam!

I'm not sure about the guy's outfit, either. I bet he's some squire who just turned up for the picture, took the fame and glory, then left everybody else to get on with the work, while he buggered off on a shooting jolly. 'What ho, pip-pip, good luck, chaps!' He was just there for a portrait opportunity.

ELI WHITNEY'S COTTON GIN (1793)

'Gin' was short for 'engine', which is a good thing, as you would not want to be operating this pissed. I bet some people loved that machine. Those who had to separate the cotton fibres from the seeds by hand before it came along will have been so chuffed.

I love the simplicity of the machine, the way the picture just shows you what it does. That's how I want things to be advertised to me: if I'm buying it, I don't care about exactly how the technology works. I just want to know where it goes in, where it comes out, and what happens to it in between.

THE HEATHCOTE STEAM PLOUGH (1837)

Not only would I not want to use that, I would not want to go anywhere near that. It looks as if it's going to rip your arms and legs off, however well you're driving it. If you go within fifteen feet of that, it's going to kill you. They should set up an exclusion zone around it, with signs saying: 'CAUTION: YOU WILL DIE'.

ANNA BALDWIN'S MILKING MACHINE (1879)

It must have been brilliant when this came along. It must have looked like something you'd put on your head to go into the ocean, but what a difference it must have made. I've milked a cow by hand. It isn't half f***ing hard work. By the end of it, my hands felt like they were going to fall off, and I had no movement in my wrists for two days.

No surprise this machine was invented by a woman. Women did most of the milking in those days. People don't realize how much farm work women did then, and still do – and how tough they are. Women farmers are amazing. Farming has been a genuinely equal opportunity occupation for centuries.

There are some jobs at which women are much better. Lambing, for example. Generally, women have smaller hands and a gentler touch. Me trying to do something that fiddly is virtually impossible with my great big shaky hands, but my other half is great at it. If you've got a lovely woman who's going to help you lamb the sheep, make the most of it. She can get on with that, and you can just heave the bales around.

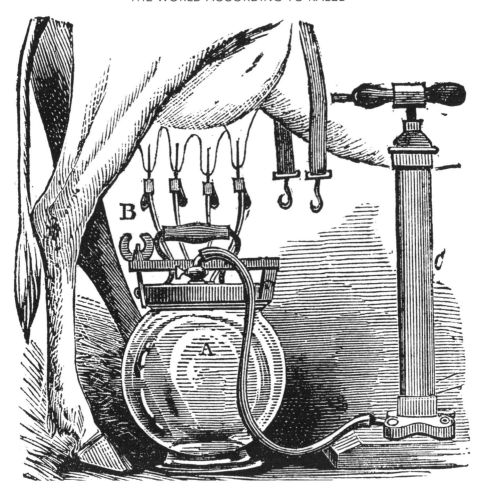

THRESHING MACHINE (1881)

Again, this is something you couldn't pay me to go within ten yards of. It's just belts and cogs, and people have to stand on top of it. Imagine getting your arm stuck in this when it's working, and the horses are going round and round. You're not going to be able to stop the horses in a hurry, are you? With a tractor,

you can just turn it off. With this thing, by the time the horse gets the message, your arm will be long gone. It's not so much a threshing machine as a killing machine. No wonder that bloke on the left has got that look on his face. He's counting the seconds until it's his turn to go on the other end, and survive until bedtime.

HORNSBY HEAVY OIL MILITARY TRACTOR (1903)

Wow, turn the page and look at that – the engineering is absolutely fantastic. I like how you can see everything, right out in the open. If I have a breakdown on my tractor, it's hard work to get into it, and so it's going to cost me a lot of money as so many of the parts are hidden or electrical. Here, you can easily see what's broken, and there's enough space for you to get at it.

Still, look at the size of it! It's like a twelve-tonne bus going around a field. It's not going to do the soil compaction much good, is it? I know a farmer who's still got some of this stuff. He spends all day in the workshop either looking at it or fixing it. That tells me that it must have broken down a lot, because he's still to this day taking it apart and putting it back together. It's a miracle his marriage hasn't broken down as well.

HORNSBY CHAIN TRACKED TRACTOR (1907)

Now that is a weapon. A next-level weapon. No wonder they've got an old-school squaddie driving it. Imagine if it fell into the wrong hands. Good thing Hitler never got a go on it.

I'd love to have seen that working. And the sound of it. It's probably not very good for the environment, but it would be worth it just to see and hear it go.

Chapter Six

Sport

'Piece of piss when you've faced the mighty Oxford City All Stars.'

When it comes to sport, everyone thinks I'm a bit … different. OK, not just when it comes to sport. I'll be honest, they've been telling me all my life, 'Kaleb, you're a bit different.' Sometimes they use other, shorter words.

It's fine by me. I've never had a problem with being different, and it's served me all right so far. Everybody thinks, you're a farmer, so you're a rugby man. You've got to love rugby. Well, I don't. I hate rugby. Rugby hurts, and you can't get to work the next day.

'Dunno what you mean, feller, we're a bunch of pussycats.'

I was made to play rugby in school. My schooldays are a blur, partly because I left school very young, then had to go back to do my GCSEs. But mainly they're a blur because of rugby. Rugby at school is basically compulsory physical violence. The only good thing about it is that you get to find out if you're the one kid everybody hates. If you end up at the bottom of the dogpile every single PE lesson, that's probably you. And now I think about it, that's not a very good thing at all.

'This is a level of insight I could really do without, on a daily basis.'

I'm a football man, me. If I go and play rugby, and I break my arm, I'm not much good on the farm. But if I go and play football, I can just dive to make it look like I'm injured. That's a trick I learned from the pros. They get subbed off, they get their money, they're right as rain the next day.

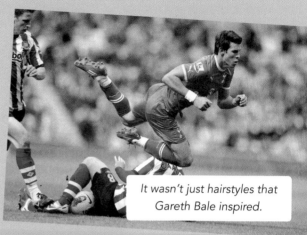

It wasn't just hairstyles that Gareth Bale inspired.

I've been into football since I was ten or eleven. My gramps got me into it. I used to play goalkeeper for Chipping Norton Town Swifts. I was a pretty good keeper. I got into the first team, and made the *Chipping Norton News* quite a bit.

Then I got injured. Genuinely and seriously. I was playing on a bad pitch and I got my foot caught in a pothole while diving for a ball. I tore the ligaments in my leg. That's when I got into farming and found that I enjoyed it more. On a Sunday morning, when everybody else was playing football, I realized I'd much rather be scraping out the cows.

Why not both?

'Just give us a bit of magic sponge, lads, and I reckon I can stay on for the second half.'

Years later I heard about a guy called Bert Trautmann, a goalie who played through an FA Cup final with a broken neck. He was a Manchester City player, and I'm a Man United fan, so I'd like to give him credit for that – but I won't.

Just because I like sport doesn't mean I have to be sporting.

Supporting United was something else I got from my gramps. That's how it works: when you're a kid you support whoever your dad does – or in my case, your grandad. If I hadn't, I think I'd have been kicked out of the family.

I don't know how he became a United fan himself; whether he got it from his own dad or he chose it. He can't have been a glory hunter, because it happened in the seventies, and they were rubbish then. Even worse than they are now, which is saying something. Maybe that was why. Country people can be bloody-minded like that. 'Oh, they've been relegated? Right, they're the team for me, then.'

Or maybe Grandad just liked a good pitch invasion.

But in my case, when I became a United fan they were the best team in the world, winning everything under Sir Alex Ferguson. So I've seen the good times, and since then I've seen some pretty sh*t times – and I mean times that make scraping out cowsh*t look pretty. I suppose that's what it's all about, taking the rough with the smooth. These days I sometimes joke that I'm a Newcastle fan born and bred, and have been ever since they got their hands on a hundred billion quid. But I don't mean it. Once you've got your football club, you don't ever change. Even when they seem to be doing everything they can think of to make you want to.

'Say what you like about being a United fan, it's better than a poke in the eye with a sharp stick – Oh.'

My favourite sportspeople have all been Manchester United players. Like Nemanja Vidić, the defender. He reminds me of the kind of kid you would always try to avoid playing rugby against in school. Or always try to avoid full stop. If he's got the ball you just stand there and watch him. You don't want to touch him because he'll throw you about thirty feet. He's like a bull running towards you. You do not want to get in the way.

The fans used to sing a version of that song 'Monster' at him: 'What's that coming over the hill? Is it Nemanja? Is it Nemanja?' Or when he came up for a corner, they'd do the theme from *Jaws*: 'Vi ... dić. Vi ... dić. Vidić Vidić Vidić Vidić Vidić Vidić ...'

'Well, this has been jolly! Let me boot you into the stands again sometime soon.'

Ronaldo was another favourite of mine, but he gets enough publicity as it is. If I'm honest, what I like about football the most is, it's a social thing. A chance to get your mates round and have a cider. That's why a player I really liked was Nani. The main reason being, it would involve my nan in the football chat. If somebody mentioned Nani, she'd turn around and say, 'Yes? What is it, love? Do you want a cup of tea?' 'Oh, we were talking football.' 'Oh, yes, how did you get on at the weekend?' 'Oh, really good thanks, Nan. Ji-Sung Park scored two, Ronaldo scored one and Rooney scored two.' She'd have no idea what you were talking about, but it was nice to make her part of it.

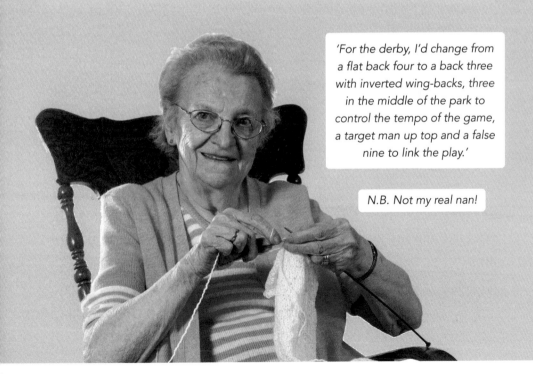

'For the derby, I'd change from a flat back four to a back three with inverted wing-backs, three in the middle of the park to control the tempo of the game, a target man up top and a false nine to link the play.'

N.B. Not my real nan!

Football's at the top when it comes to sports I like. I love playing cricket, although I hate watching it. And I hated watching it before everybody else in England hated watching it too.

'Yeah, but now we've given them a reason.'

I like a game of rounders with my friends, too. But when it comes to playing sport with your mates, nothing beats 'bum slap'. A lot of us young farmers play it. It's a very old-fashioned game.

You've got to score a goal on the volley. But if it's not a volley, or if the keeper saves it, then you have to go in goal. For example, if you've got five people playing, and I'm in goal, and I let five goals in, then I come out of goal.

You get ten or eleven lives. When you're on your last life, you get an extra life called a 'doggy life'. Then the goalkeeper has to bend over in the goal, and you get to take a shot, and if you hit him on the arse you can have another go, but if you miss, you have to go and stand next to him.

It's all very straightforward and easy to understand, but for some reason, whenever I explain the rules to anyone who isn't one of my

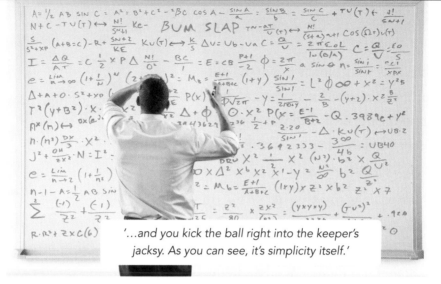

'...and you kick the ball right into the keeper's jacksy. As you can see, it's simplicity itself.'

young farmer mates, they get confused and their eyes go glassy and they try to change the subject.

I think bum slap is in the tradition of a lot of country sports, like cheese rolling, where everybody chases a wheel of Double Gloucester down a really steep hill and ends up in hospital. Or the sort of football games they had in the old days where everybody ended up in hospital. Or would have done if they'd had hospitals.

Most countryside sports seem to involve complete mayhem and everybody getting hurt. All the spectators find it very entertaining. But I'm not sure why the competitors do it. Perhaps it's for the recognition. The trouble is, nobody's going to recognize you once you've lost an arm and your face is on backwards.

EARLY DAYS OF SPORT
ANCIENT FOOTBALL IN THE 14TH CENTURY.

'Keeper's!'

I'd rather walk slowly down the hill and eat the cheese when I get there, instead of breaking my neck and wasting the cheese. I've never been that keen on dangerous countryside sports myself.

There was the one time when air rifle practice became dangerous. It started off as target practice – we shot a couple of tin cans and had a laugh – but then I ended up being the target and got shot in the leg. Then again, my legs are usually a particular combination of white and red that looks like a tin of budget baked beans, so you can understand the confusion.

'You've caught a bit of sun on your legs there, mate.'

And there's ferret racing, which may not sound so bad, but that's only if you haven't met ferrets. They are the most vicious animals

I've ever encountered in my entire life. I nearly lost my finger to one. I'm sure if your ferret was losing the race, you would just need to dangle a finger in front of them. That would get them moving. It's like a rabbit to a greyhound, only you're the rabbit. And if you don't get out of the way in time, I'm telling you, get used to making do with nine fingers. Ferret racing is a sport the way Russian roulette is a sport. And I'm not about to take that up in a hurry.

Look at it, though. A severed finger wouldn't melt in its mouth.

KALEB'S COTSWOLDS WORKOUT: GETTING FIT THE FARMING WAY

BEDDING THE COWS DOWN

This'll start you off nicely. Chucking the bales around will give you a good warm-up stretch, get the heart rate up and keep the circulation moving. You don't want to go in cold, and nor do the cows.

'I like to mooove it, mooove it!'

'Drat and botheration, this is a blessedly tricky task!'

CHANGING TRACTOR WHEELS

There's a story that Hercules – you know, the guy from the Disney film – would go out into the pasture as a kid every day and lift the same bull calf up above his head. So, by the time he was an adult, he was lifting up a fully grown bull. Which isn't a bad farming workout, I admit, but bench-pressing a bull has got nothing on changing tractor wheels.

Tractor wheels are ridiculously heavy. You may need to build up to them over time by changing the wheels on a truck before you make the step up to a tractor. And if you really want to put on lean, mean muscle, try fitting the tyres to the hubs as well. You'll burn off pounds of body fat by swearing alone.

GO AND GET YOUR SHEEP BACK IN

Or rather, go and see if you *can* get your sheep back in. That's a workout and a half. It's like high-intensity interval training, only with woolly suicidal idiots sent from the depths of hell to torment you. Show me a personal trainer who can match that for motivation. If you want to get your cardio up, just buy some sheep.

Going for gold in the sheeplechase.

This is supposed to be cruel to the pigs, but frankly they seem to be getting the better end of the deal.

PIG WRESTLING

You'll want to get greased up for this one. This'll do your back muscles like nothing else. Either that, or it'll do your back.

DRIVING IN FENCE POSTS

You're going to be shaking for about three days after this. I don't even know the real name for the thing you use, or whether it has one. Post-knocker? Fence-rammer? I've always just known it as the 'man-killer'. When an older farmer used to say, 'Go and get the man-killer', I'd know exactly what he meant. Trust me, it'll live up to its name. But if you survive it, you'll be ready for the Olympics.

One man down, two to go.

Chapter Seven

Culture

'Are you even listening right now? What did I just say?'

I'm not going to pretend that I'm an aficionado of the arts. Or even that I knew what 'aficionado' meant until around five minutes ago, when I asked my other half, 'What's a word for somebody who likes books and things like that?' It wasn't the first word she came up with, but it was the first printable one.

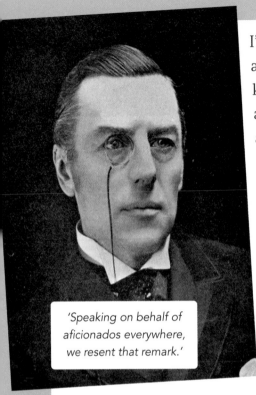

'Speaking on behalf of aficionados everywhere, we resent that remark.'

I have to admit, it's quite satisfying to think that the first book I ever own will also be written by me. In your face, culture snobs. Some people spend their whole lives trying to get a book published, and I do sympathize with them, but before they get too envious, they should stop and ask themselves if they'd like to have been doing eighty-hour weeks on the farm since they were thirteen. I've worked hard for this, just in a different way.

'I realize my mistake now. I should have bought a tractor.'

Books were never my thing at school, because I didn't have the concentration to read the whole thing. And a lot of the time I'd be off working anyway. But sometimes

Frankly, it doesn't end well. Or start well. Or go very well in between.

I'd hear they were going to show us a film of the book we were supposed to be studying. So I'd always make sure to go in on that day. That's how I found out what happens at the end of *Of Mice and Men*, which I was enjoying for a while, because it was all about farm work. I'm not going to spoil the ending for anyone, but the ending certainly spoiled it for me.

It didn't have any mice in it, either, although the big lad was always talking about how much he liked rabbits. I can only guess he'd never actually had to deal with rabbits on a farm. They're a pain in the arse. They eat so much of your crops. They're just fluffy pigeons with funny ears. People get sentimental about them, but I bloody hate bloody rabbits to the maximum. So here's another book you can keep: *Watership Down*. When us country people watch the film of that, we're cheering on the diggers and bulldozers and cats, and anything else that'll keep the rabbits out of our fields.

Bri-i-ight eyes, yummy for dinner …

Winnie-the-Pooh and Christopher Robin

For services to parents suffering from sleep deprivation.

I do know a bit about a few other books with a countryside theme. For example, I know the Winnie-the-Pooh books are set in a hundred-acre wood. But I don't think that's why they're popular. Those are bestsellers because parents know they can buy them and their kids will be quiet for a bit. I can see why the bear and the boy ended up on a stamp. The guy who wrote those books is one hundred per cent a national hero.

And I've heard of *The Wind in the Willows*, which is an adventure with a rat, a mole and a toad. Honestly, I can't think of anyone worse to go on an adventure with. There's also a lot in it about the great god Pan, which nobody seems to mention, and I can understand why. Which is more than I understand about who the great god Pan is or what he's doing in a kids' book.

'Hey, baby – how you doin?'

All that pagan stuff is fine if you don't actually live and work in the country. I can see why people might like the idea of dancing through the woods in the nuddy like a lunatic, blowing into pan flutes, then going home to where it's all pavements and plumbing, and well you might. I believe more in dinosaurs and evolution, myself. We were monkeys once and now we're not. Not all the time, anyway. I have my moments, I grant you.

'A credible hypothesis, and one with which I find myself in accord.'

Art is another thing I don't really connect with. When I look at a painting, I don't understand what it's trying to tell me. I don't really get the point, other than somebody's going to stick it up in a gallery and convince somebody else to pay mega-money for it. But why would you ever want to spend your hard-earned cash on some splodges?

'Not gonna lie, it's a chilli pepper with a head and feet – yours for three million.'

There's one guy I know who's an art dealer. And he smokes a lot of weed. I cannot help but think that those two facts are related. I reckon what happens is he goes out there, looks at a bit of art, smokes a joint, and says, 'That's amazing!' And that's how he determines if something's good or not.

'Good afternoon, I represent a prominent international dealership and ... oh, wow, do you want to sell me that light fitting?'

I've got four pictures hanging in my house. One is my *Wolf of Witney* (see p.169). One is of my dog, the one we lost last year. The third one is of me and my other half with the same dog. The fourth one is a picture of a sheep standing at the edge of a cliff. Something about that one speaks to me. It's a lovely scene in the background, but I don't think that's why I like it. I think it's because I'm hoping the sheep will go over the edge. One day I'll walk into the room and it'll just be a picture of a cliff, and I'll think, 'Yes! At last! Justice is served!'

Ars longa, vita brevis, ovum horribilis.

I've been told that if I don't understand traditional painting, I should try conceptual art. Which, as far as I can tell, is art that isn't really there, or art that isn't really art. And when I say that, people go, 'Yes, that's the point.' Well, OK. It's your money. It all began when some French guy bought a urinal, signed it and put it in an exhibition. I'm amazed nobody actually tried to use

Oh, thank God. I've been here hours, and I'm busting.

it. Although perhaps that would have increased its value, as an interactive piece.

That was more than a hundred years ago, but I bet things haven't moved on an awful lot. Wasn't there that woman who messed up her bed, threw a whole load of towels onto it and sold it for millions of pounds?

Now, music is something that I'm much more into. What I really like is country music. Which I suppose makes sense, because I'm a country person, if you haven't realized that by now. And I'm enough of a connoisseur (another word my missus taught me) to know that I like English country music even more than American.

They are very different. Everything in America is bigger. It's all about trucks and prairies and cowboy hats. If you tried pulling any of that kind of thing around here you'd get mugged off well and proper in five seconds flat.

Mate, the festival isn't even on this year.

What if Status Quo, but agricultural?

The Americans have got Hank Williams, we've got the Wurzels. In American country songs, they're always singing about freight trains, long-distance lorry drivers, pickup trucks. The Wurzels sing about combine harvesters. That's much more relevant to my life. People think the Wurzels are a joke, and perhaps they are, but in the right way. I bloody love them. I've been to see them and they're brilliant.

Not that there's anything wrong with songs about freight trains. In fact, one of my favourite songs ever is called 'Freight Train', by a guy named Alan Jackson. It's an absolute banger. Especially in a tractor. You can't beat a proper tractor playlist on your headphones. Although these days you can get a tractor with a bass speaker fitted behind the seat. I'm not saying that's what Alan Jackson would want you to do, but I'm not saying it isn't, either.

'Howdy, I'm Alan Jackson, and I probably approve of you listening to my music on a bass-enhanced tractor, or anywhere else you darn well please.'

Also, in American country music, some fella is always getting left by somebody or something – his missus, his dog, his truck. So he goes to a bar and starts drinking, and that's a bad thing. In the British equivalent, drinking is great.

The Wurzels sing about drinking cider. That's just about the only thing young farmers do when they're not actually farming. And it's not a terrible, depressing thing. Drinking cider with your mates is fantastic. I'd recommend it to anybody. Perhaps that's the trouble with all those saloons and honky-tonks. Too much whisky, not enough cider. No wonder they're all poor miserable sods.

Television is a funny one for me because I didn't watch much TV before I actually started appearing on TV. And now, making TV has ruined me for watching it. I can't really enjoy it, because I can see through it. I'll be thinking, they should have cut that there, or I can see how they fixed that in the edit. I can't do what the director would call 'suspending my disbelief'. It's a bit like how most people don't have to think about where their food comes from, but if you're a farmer, you know. You literally know how the sausage is made – and you know what to avoid.

'Way I figure it, feller, our'n problem here is they don't serve no scrumpy round these parts.'

Wait, haven't we seen that somewhere before?

Yes! We knew it!

127

KALEB'S GUIDE TO THE BEST IN CULTURE

FAVOURITE BOOK

I'll go for *Animal Farm* by George Orwell. It was always going to be this one, wasn't it? For the title alone. It's about a group of farm animals who get together and drive away the farmer, which is never going to happen, but would be pretty cool if it did.

You probably think I'd be on the side of the farmer, but not here, because the farmer is evil. I don't like farmers who give us a bad name, when it's farmers who look after the environment so much more than anyone else out there.

The pigs are the ringleaders in this one. They're the brains behind the rebellion. And that makes sense. Pigs are very clever animals. They always poo in the same corner of their pen. And if they don't like a bit of food they'll take it to that corner and drop it there rather than leaving it in their trough. They'll cut out the middle-pig, you could say. They build their own draught excluders out of straw to keep their sleeping area warm. So I can definitely see pigs leading the rebellion.

What doesn't ring true for me is that the sheep in this book are all very obedient and always do what they're told. Which is a bigger load of sh*t than the one in the corner of a pig pen. When

I was looking this book up, I found out that it's really about the overthrow of capitalism and the failure of communism – an allegory, apparently. I don't know anything about the political part, but I do know that there aren't any giant predatory prehistoric aquatic lizards in it. I'd definitely have remembered that. It's totally my kind of thing.

'We're the ones with the broad snouts. The ones with the narrow snouts are called caricatures.'

FAVOURITE TV SHOW

I can tell you what my least favourite TV show is, and that's *Countryfile*. It should be called *Hobby Farming Style*. I don't really have a favourite TV show, apart from the one I'm in, although I love anything with David Attenborough and animals.

Having said that, when I was a dairy farmer, I would go and have lunch in the house, and there was always a show on about a vicar who used to go around solving crimes. *Father Brown*, it was called. I liked that, because it was filmed in the countryside. A shame he never sorted out whoever's behind *Countryfile*, but you can't have everything.

'This week I 'ave been mostly wearing cassocks.'

FAVOURITE PAINTING

There is one painting I really like, and that's *The Hay Wain* by John Constable, who I'm guessing also used to go around the countryside solving crimes. He must have done the art in his spare time.

This one's really nice. As I said, when it comes to splodges on the wall that people pay millions of pounds for, I have no idea what they're on about. I wouldn't put it in my house or anything, but this is a picture I can really see into. I can imagine myself driving that cart back from the field after a hard day, thinking to myself that I'll cool the horses off by going through the river. Anything to do with the history of farming, I can enjoy. It's as if it's me in there, doing whatever they're doing, and I like that.

We like that, too. It's a lovely picture and a lovely thought.

Chapter Eight

Health & Safety

'That's all very well, but I could use a hat at least.'

'This is a sub-optimal situation.'

Health and safety on farms is a massive issue. The occupational death rate for farmers is so much higher than the average, it's not even funny. You'd probably be better off working on an oil rig, or in the army, or even in the army on an oil rig. People should be more aware of the serious risks farmers undergo every day to supply them with their food.

One big problem is that the work we do on farms – every single part of it – is just so dangerous by its very nature. The other big problem is: farmers.

I'll give you an example. You're not supposed to use the telehandler, ever, without a seatbelt on. 'Clunk click every trip', as the old public information film had it. Although it tells you something about the seventies that we were expected to take safety advice from Jimmy Savile.

Also, you're supposed to apply the handbrake between every movement. Do you think most farmers, or any farmers, actually do all those things? If I'm bale-carting in the field and I stop and put the handbrake on each time, it's going to take me forever. I'm never going to get to the pub.

'Fasten that belt, lift that bale ...'

The two biggest factors undermining health and safety on the farm are the pub and the rain. You're always trying to get to the first one, and beat the second one.

But even if you leave the King's Arms and the weather out of it, the other problem is that farmers are always ridiculously busy, always in a hurry. Not only will farmers do insane things that nobody else would do, and do them at a pace that nobody else could do, they'll also keep doing them no matter what. If a farmer loses an arm in a thresher, they'll put a plaster over it and be back at work the next day.

'Yeah, it stings a bit, but let's stick some Dettol on it and I'll get back to the combine.'

I've been in a few life-threatening situations in my time. Once, I was calving a cow and I broke three ribs. The bull was next door and tried to come through. He caught me and flipped me up in the air and I landed on a gate.

I've been stuck in a ring feeder for an hour and a half, hiding from another bull who decided he didn't like me that day. Brian, his name was, a Brown Swiss – and I've always thought that was a very Brian-the-Brown-Swiss-bull kind of thing to do.

'U wot m8?'

Yeah, that thing will aerate you, all right.

Us farmers can be our own worst enemies. I've nearly rolled a tractor, trying to lift too many silage bales at once – just because I thought I'd do one trip instead of two. Save a little bit of time. I almost turned over on a bank while aerating near Long Compton.

Another time, I was going downhill in low gear and the aerator I was pulling overtook me – and if you've ever seen an aerator, that is not something you want anywhere other than exactly where it belongs: a nice safe distance behind you.

I had to think quickly and ask myself, how am I going to get out of this situation? I tried to steer inside it, but that didn't happen, so I turned uphill and let the clutch go, and ended up doing a wheelie.

Because of that, I realized the wheelie position was great, because I could get lovely straight lines on the aerator. Drop the clutch, pop a tractor wheelie, and you can't turn the wheel even if you want to. Result!

Of the two possible outcomes – experiencing a horrible grisly death, or learning a really smart way to do something – I got the right one.

You can keep your lowriders, this is the business.

At the time, I was only an apprentice and I didn't have a lot of experience. But the best way to get experience is to go out there and do it. Of course, there's always the question of whether you'll survive it. But if you do, it makes a great story for the pub.

This is why the accident book plays such a central role in farmers' lives. Any little accident you have on the farm, you have to go and put in the book. The majority of the time it's for your farm assurance – the certification process for agricultural products. The authorities may think your farm is absolutely amazing and you've got the best produce, but they still want to come in and check on you. So they do an inspection, once or maybe twice a year. They come in and look at your accident book.

If I cut the end of my finger really deeply, I've got to go and write about it in the accident book. What happened? 'I was being an idiot and trapped my finger in the gate.' Who witnessed this? 'Nobody. Unless you count some cows.' What did you do to make sure it was looked after? Normally, I'd write, 'I sought advice from a first-aider' or 'I went to the doctor's', or 'the hospital'. But the truth is, I've usually washed it under a tap, put a glove on it, kept milking the cows, watched the glove fill up with blood, then just binned it.

The accident book is covered in blood spatter and gore. I once cut my right thumb so deeply it caught the nerve. I still can't feel anything in that thumb. But I had to write in the accident book with it anyway, because I'm right-handed. The mess I made, I might as well have written it in actual blood – just chopped off the end of my index finger and done it with that.

You've got to put everything in this book. It knows your life better than anything else: where you've been, what you did there, and what tried to kill you. It's like a literal Doomsday Book.

'OK, so you just write, "Fell onto aerator while popping a tractor wheelie" here, then put your name there ...'

If they ever made a film out of the accident book, it would look like one of those serial killer movies. Or a horror film like *Final Destination*, full of all these ridiculous ways to die. I think it would be even better as a TV show: to go around every farm and look in the accident book, then re-enact the worst or the stupidest ones. You could call it *When Farms Attack*.

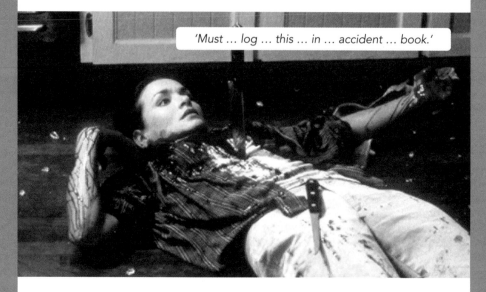

'Must … log … this … in … accident … book.'

The truth is, though, that things are much, much better than in the old days. And by the old days, I mean, not all that long ago. It took a long time for farming to catch up with other industries in terms of health and safety. They didn't even start legislating for it until around the fifties. So there might be farmers alive today who remember when there were no health and safety rules at all. Although that does seem unlikely. I've got no idea how they'd have made it into their twenties, let alone their eighties.

'I remember when it was all fields around here and … wait. All right, then, I remember when I had fingers.'

You only have to look at loadalls, which are the same thing as telehandlers, one of the most commonly used pieces of farm machinery. It used to be that if you wanted to get on one of these and get lifted up in the air, you'd be standing in a giant bucket. Obviously, that isn't ideal. So now you've got these things called man-cages …

We looked up 'man-cage' on the Internet, but we couldn't print any of the pictures and now we're on some kind of watchlist.

… and it's legal to go up in them. Back in the day, all that equipment wasn't checked on a regular basis – or on any basis at all. If there was a leaky hosepipe on the rig letting the fluid out, all of a sudden it could burst and drop you. Nowadays, you have a loadall inspection.

That's just one example. You also get training these days, which never used to happen. You get PPE – personal protective equipment. Imagine being sat on a combine harvester all day, with no cab, and all the dust coming into your lungs. That's not going to be good for you.

These days, the people who make the machines actually think of the person who's going to drive them. In the old days, they didn't care. Their attitude was: as long as the machine works, as long as we've got our corn in, we don't give a stuff if you die.

You've still got a problem, which is that farmers are tight. You can get a proper man-cage which costs £3,000 and not even a bear could break into it. Or you can get a cage for half that price which just has a bit of a barrier around it. I know for a damn fact which one a farmer is going to choose. And I can assure you, the bear is definitely getting into that one. Or you're falling out of it.

Farmers get grumpy about the Health and Safety Executive because the HSE makes them do things that cost time and money. They're always complaining because they don't want to do this or they don't want to do that. But secretly, in their heads, they know it's the right thing, so they go ahead and do it all anyway. They're much more involved in getting the proper training and protection for the young people, the apprentices, working on their farms. And they don't want the backlash, either. If I have an accident on their farm, then it's their fault if they haven't given me the right training, and they're going to get sued. So they may not like to admit it out loud, but they accept the HSE will save people's lives and help people in the industry. And that it'll save their own arses from the consequences of things going wrong.

But what if the bear falls out of it?

'We are not so very different, you and me.'

KALEB'S GUIDE TO THE BIGGEST DANGERS HE FACES ON THE FARM

CONTACT WITH MACHINERY

This is what the HSE calls near-fatal accidents, which is a nice way of putting it. If you break down when you're out on your own in a field, you've got to try to fix the problem yourself. You get in the cab, turn the machine on, climb down to make sure it's working properly, think, 'Oh, I might just lift that little bit of metal to check –' Then WHAM!

And nobody else is there to help you. I can be out in the fields eighteen hours a day and nobody knows where I am. Farmers are a bit like those cartoon characters who look down the barrel of a gun to see why it isn't working. There's definitely a touch of Wile E. Coyote about us.

'Oh, boy, the HSE will not be happy about this.'

CONTACT WITH ELECTRICITY

I'm big on avoiding this one. I do a lot of training on it, because I've heard real horror stories. Part of the problem is: farmers again. They think they can do everything: 'I'm a carpenter. I'm a builder. I'm an electrician.' You're definitely not an electrician, mate, and one day you're going to find that out the hard way.

DANGER OF DEATH BY 3,000 VOLTS (INCLUDES FARMERS)

Maybe it would be simplest just to attach one of these to the tractor. Or to the farmer.

The most common accident occurs when a famer goes into a field on a tractor and there are overhead power lines. He tips the trailer and catches one of the lines, and without thinking he tries to get out. But the tyres are insulating him, and as soon as his foot touches the ground with some other part of him still touching the tractor, he becomes the earth, and ... ZAP!

What you're meant to do is bunny hop. You get to the top step, where you're still protected by the rubber, and you jump as far away as you can. You're probably going to break a leg, but breaking a leg is better than being electrocuted, trust me.

ZOONOSES

These aren't the snouts on animals you see at the zoo, which is what I first thought. They're diseases that make the jump from animals to humans. For instance, I've caught ringworm from a cow. I'm sure there must be other, better known, more topical examples, but I can't think of one off the top of my head.

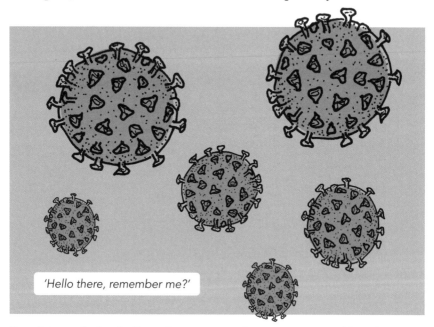

'Hello there, remember me?'

Speaking of which, I've also managed to inject myself with various things that a cow should have and I shouldn't. I've missed the cow with the needle and hit myself instead. Also sheep. I Heptavac'd myself while vaccinating sheep. I went weird for a little bit after that one. Still, I haven't caught any sheep diseases, so I'd like to think it helped.

DIRECT SUNLIGHT

This a specific problem for me, given that farming mostly
takes place in direct sunlight (and if it doesn't, you're doing it
wrong). My skin has two colour settings: albino and tomato. Five
seconds in the sun and I'm a weird combination of both. I need
factor five million sunblock, the kind that normal people would
use if they were actually going to land on the surface of the sun.

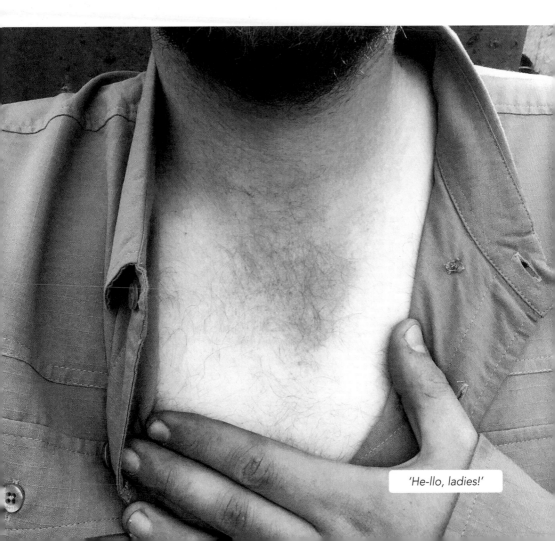

'He-llo, ladies!'

JEREMY

(Enough said.)

'Moi?'

Chapter Nine

Philosophy

'Only the dead have seen the end of sheep.'

Falling leaves: not always that quiet.

I have to admit, I don't spend a lot of time thinking about the meaning of life. I'm more interested in doing things, and by 'things' I obviously mean things that involve me driving a tractor. Sometimes I end up in conversations in the pub where my friends get all deep and meaningful and ask questions like, 'If a tree falls in the forest and nobody hears it, does it make a sound?' To which the answer is, 'Yes, of course it does.' Because that's how sound works. I may not have been top of the physics class at school, or even in the physics class at school, but I know what bloody soundwaves are.

I also know that if you make a cock-up in the middle of nowhere, somebody will see it. And everybody will find out about it. Guaranteed. Probably because I'll be the one who sees it, and I'll tell everybody. That's what we all do in the countryside. There's nothing quite as much fun as seeing a mistake somebody thinks they've got away with and broadcasting it to the whole village. And I know why cows are always looking over the fence. It's because the grass is always greener on the other side. Literally.

Getting to it is another matter, though.

But now I'm a man of letters – all twenty-six of them – I thought it would be a good idea to find out a bit about philosophy so that I can hold my own when it comes to talking about why we (and, in particular, sheep) exist.

So, here's the stuff I've learned that I can actually relate to. People who are more interested in doing things than thinking about things are called 'existentialists'. They were mainly French people from the black-and-white days, and they said things like 'to do is to be', which sounds up my lane. It could be that they were actually French farmers, not just any old French people. That would explain it.

When you have to deliver a lecture at the Sorbonne at 11am, but you need to get the sheep dipped by 9am.

'I'm pink, therefore I'm lamb.'

It seems that if you want to be a philosopher you either have to be old and French, very old and German, or unbelievably old and Greek. At first, I thought that ruled me out on all counts, which was a relief. I could just let them get on with it and try to work out what they're on about when I have a bit of spare time – which, what with me being a farmer, is almost never.

But, according to one of the old French guys, René Descartes, if I don't do the thinking, I won't exist. 'I think, therefore I am,' he said. Which I suppose is true. But, on the other hand, sheep don't think, and they still exist. Why? Nobody knows, apart from the fact that they taste really good.

'I don't think the grass is greener, I'm just an idiot.'

Perhaps this is one for William of Ockham, who managed to be a philosopher, even though he was English, and was famous for having a razor, which you'd think would have made him a barber.

Occam's Razor – they changed the spelling of his name for it, don't ask me why – is a principle that says that 'entities should not be multiplied beyond their necessity'. Which, when you translate it from the original gibberish, means something like: don't look for a complicated explanation when a simple one will do the trick.

For instance: why do sheep get stuck in fences? Is it some complex phenomenon of animal behaviour? No. It's because they're suicidal morons. Which explains pretty much everything else they do, too.

It all gets very complicated with the French philosophers after that. There's something called 'deconstruction', which is usually what's happened when Jeremy's been trying to fix something.

It's related to 'post-structuralism', which is usually what's happened after Jeremy's tried to build something. In these philosophies, meaning is constantly fluctuating and nothing is stable, which is one hundred per cent like working for Jeremy under any circumstances. Even the stables aren't stable.

Apparently, the post-structuralists all had it in for some poor sod named Lévi-Strauss, which is a bit harsh on somebody who only ever wanted to make a decent pair of jeans.

'Just wait 'til I strip down to my grundies in a launderette to a Motown tune, that'll shut 'em all up.'

I'm more interested in a guy called Augustine, who's one of the fathers of Western philosophy. But that's not why I'm interested in him. It's more that he was probably the father of a whole lot more besides. He was a proper randy old bugger. He had a famous prayer: 'Grant me chastity and continence, but not yet.' We all know a guy like that. They used to say that he'd have shagged a snake if he could have held it down for long enough. They take the piss out of farmers and sheep, but that's way worse. And he was a bishop! And then he got made a saint!

Funny how that happens – all the posh randy buggers end up getting titles, and getting promoted. I'm not making any comment on how it happens, but they always do. I wonder if it goes right to the top.

'Sorry, you are mistaken, I am in fact the late and much loved actor Warren Clarke – please do not drag me into these shenanigans.'

The other thing about Augustine, and I'm not making this up, is he came from a place called Hippo. Honestly, if all philosophers were this mad, I'd have started finding out about them long ago.

Anyway, I've got sidetracked. Let's go back to the big philosophical questions about the world – the ones I get, anyway.

The Ancient Greeks felt that if something wasn't visible, it didn't exist. There was this guy called Pyrrho of Elis, who was the first sceptic – or so I'm told, but I'm not sure if it's true. He refused to believe anything at all until physical reality made it undeniable.

He's got a point. Seeing is believing. I'm like that. Especially with farmers, who are very inclined to tell tall tales. After harvest, they'll come into the pub and say, 'Yeah, we got fifteen tonnes of wheat to the acre.' You know for a damn fact it's bullsh*t – and not even that much bullsh*t could help you grow fifteen tonnes, when three tonnes is the normal amount.

I know what they're up to because I do it as well. Wheat is to farmers what the size of fish is to fishermen. 'Oh, we caught a thirty-five-pound carp.' And it turns out it weighed only about three. So old Pyrrho was right about the whole lot of us.

'You grow fifteen tonnes, what do you get? The piss taken out of you in the pub, mainly.'

Behold! The Platonic ideal. Now, quick, hide it behind the barn before anybody else sees it.

Then there was this other theory I came across that says that – hang on, let me quote from the Internet – 'all things on Earth are merely inferior and unsatisfactory copies of unattainable ideals'.

Maybe somewhere there's a version of that sentence that would make sense to me, but not on this planet. A lot of philosophy reminds me of religion – and I'm not a massive fan of that either, as people know. They come out with something that sounds like a load of nonsense but, as long as they can get one other person believing it, they're away – the whole thing starts to spread.

As far as I can make out, there was someone called Plato who believed that there's a perfect version of every single thing somewhere, but we never get to see it. We only get to see the sh*tty knock-off versions.

That is something I definitely understand. Farmers are like that with livestock. There's always a perfect bull somewhere. But you never get to see the perfect bull. Somebody's always trying to sell you the worst ones, and making sure they keep the perfect one hidden.

One other concept that I could relate to was that reality is a meeting of opposites and everything is in flux. Which, again, is very much like working for Jeremy.

But the most interesting thing is that the guy who said this – a man called Heraclitus – was eaten alive by dogs. Which is one hell of a way to go. I'm not sure there's any amount of philosophy that can prepare you for that.

Also, they must have had some pretty hard-core dogs back in Ancient Greece. Mine might lick you to death, or you could injure yourself tripping over them, but that's about as bad as it gets.

One philosophy I'm not sure about is what this one dude Epicurus said about dying. He didn't half come out with some rubbish. 'Death is nothing to us.' Speak for yourself, mate, it's definitely something to me, and I don't want any part of it for a good long time. His idea was that, as we're all made up of atoms, death only means our atoms will disperse. But I'd prefer mine to stay in exactly the same place, ta.

Mind you, I think there might be something in the idea of reincarnation. I'd like to come back as an eagle, or some sort of predator. A crocodile – they've survived for millions of years, they live for ages, they're at the top of the food chain, and they scare the sh*t out of everybody. They certainly scare the sh*t out of me.

God help me if I come back as a sheep, though – I'd be f***ed.

'I was thinking more of larks' tongues and dancing girls, but hey, you do you.'

Anyway, what I like about Epicurus is he advised enjoying all of life's pleasures as much as you can, while you can. You might as well live life to the max. And I do. My biggest pleasure is farming, and I do it every day, almost all the time.

KALEB'S FAVOURITE PHILOSOPHERS – AND HIS LEAST

LUDWIG WITTGENSTEIN

This feller spent his life trying to establish the limits of human understanding. And fair play to him, that is quite a job. He concluded that the only things we can talk about are the things we can understand. And I think that's very true. It's like me and Jeremy: it's all very well him trying to talk about farming, but he might as well be speaking in Klingon for all the sense it'll make.

It would be the same if I tried talking to him about cars, which is why I don't. Wittgenstein said, 'Whereof one cannot speak, thereof one must remain silent.' But that doesn't seem to stop anyone, does it? People have an opinion about everything these days. If you don't know something, just leave it out.

'No comment.'

DIOGENES

He founded the Cynics, who believed in rugged self-reliance. He wore rags, and he lived in a barrel to avoid paying taxes. Which I think was taking it a bit far. But I bet he was happy. I'm a firm believer that everyone should be able to live where they want, so good for him and his barrel. I'd love us to go back to everyone having an acre, and trading stuff they grow on it. Plus Diogenes stuck up two fingers to Alexander the Great, who ruled over pretty much everything then, and got away with it, and you've got to respect that. Zero f***s given.

'Don't you dogs even think about it. I know what happened to Heraclitus.'

BOETHIUS

This lad was all about being happy with what you have. I like that a lot. I may not have a farm, but I'm happy with my four acres. It's more than some people have, so I'd never look at it and think, 'That's not enough.' I'm happy with what I've got, and I've worked hard for what I've got, and I'm happy with what is to come. Jeremy may have a bigger tractor than me, but I'm happy with mine. Especially as I've got four of them, and they all do what they're meant to. Still, as long as he's happy with his. Although quite frankly, it is sh*t.

'The consolations of philosophy are many,
but you can't beat a good tractor.'

FRIEDRICH NIETZSCHE

A bit of a nutter, this boy, to be honest. His most famous saying was, 'Whatever does not kill me, makes me stronger.' That's all well and good, but when a bull charges at you and breaks three of your ribs, you don't really think that, do you? You're not saying to yourself, 'Oh, fantastic, this is gonna do me a power of good.' I mean, it didn't kill me, he got that part right, but it didn't make me f***ing stronger. It made me more scared of the f***ing things. Stupid man.

'Nein! It is you who is stupid! Also, you are weak, and ...'
rants syphilitically in German for several volumes

Chapter Ten

Celebrity

Shirt by AgriMan. Gilet by Loam & Clay. Jeans by Homme@Quidland. Wellies: model's own, thank God.

The question I get asked all the time now is, 'What's it like suddenly going from being an ordinary little farmer in Chipping Norton to being a global celebrity?' My answer is always the same: 'Weird.' But it's also been very positive so far, and I'm grateful for that.

By the way, when I say 'global', that's not me being big-headed. I mean, I can be, but not about this. I get messages from people in Germany, Australia, all over. It's usually nice comments, too. Which apparently isn't always the way when you get famous.

'Can confirm.'

'Dear Kaleb, I have a question about a window box …'

Some of them are from people who are interested in farming, but I also hear from a lot of non-farmers too, who genuinely like me for who I am and what I do and want to learn more. It's absolutely amazing. They normally start off by saying, 'I'm not a farmer, but I love to be out in the countryside like you, it's beautiful out here.' Or, 'You've really inspired me to do something, so I'm growing my own veg.'

When I was younger, I never, ever, ever imagined I might be a celebrity or even thought I might want to be one. I'm a firm believer in setting a goal, then getting there. When I was thirteen, my goal was to be my own boss by sixteen. I accomplished that.

Then my next goals were to have a successful contracting business, and I did that too, and to buy a farm, which I haven't yet, but I will. All this other stuff was just extra, along the way. And without being goal-oriented, I never would have become a celebrity. My motto is: 'Dreams don't work unless you do.' If I didn't have a mindset of doing eighteen hours a day, working my arse off to accomplish my goals, none of this would have happened. I hate it when anybody says that they're jealous of me. Why? You could do exactly what I've done.

By working hard, it turns out I bought a ticket for a lottery I didn't even know I was taking part in. I wasn't trying to be famous, I was trying to get on, and even if the celebrity part had never happened, I'd have been more than happy with being a successful farmer.

'That lucky bastard Kaleb gets all the breaks. Ooh, Bargain Hunt is on.'

I'm still the same person I was before. Or at least I like to think I am. I'm still farming, still doing the eighteen-hour days, still mucking out horses and cows, still shovelling sh*t. I don't think a lot of other celebrities are doing that.

It's definitely changed my life, though. Mainly because when I go to the shop to buy some milk, I have to do about 50,000 selfies on the way to get it, and

Well, then again …

another 50,000 to get back out of the shop – it ends up taking about an hour and twenty minutes just to get a pint of milk. And often I forget what I went in there for, because there are so many people talking to me. I'm now thinking the best thing is to buy a cow and milk it myself, rather than just look after other people's cows for them. I've never had a cow ask me for a selfie. Not yet, anyway.

Give it time.
Give it time.

I meet a lot of celebrities now, and I have no idea who they are. I only realize they're celebrities when somebody tells me. There's an event I go to each year called the Big Feastival. Alex James runs it, but I didn't have a clue who he was until I met him. They told me he was the bassist out of Blur, but then they had to explain who Blur are, and it's all still a bit of a ... well, you know. I remember something about living in a very big house in the country, which around here doesn't really narrow it down.

Also, Blur's depiction of farming life may not have been entirely reliable.

In fairness, Tom Walker looks more like a random passing farmer than most random passing farmers do.

Because I'm a celebrity now, when I'm at an event I get to go into the VIP backstage bit. The thing is, I have no idea who the other people there are. In one place, some bloke shouted a greeting at me and I waved back and walked straight past, thinking it was a fan. But my other half got very excited and said, 'That's Tom Walker, the singer!' So I went over and said hello and we had some pictures taken, but to tell the truth, I was still none the wiser. Somebody I later found out was Simon Pegg also wanted us to have some pictures taken together, so I said, 'Yeah, sure, but I don't actually know who you are.'

It seems to me like all the celebs know each other. But it's different for me – I don't recognize them because I'm on a tractor most of the day. It's always possible that other celebrities find it refreshing if I don't know who they are. I definitely would. But they probably think it's rude. As if it's me saying, 'I'm so busy now, I don't know who you are.' When the truth is, I was so busy before that I didn't know who they were, and nothing's changed there.

And for them, being a celebrity is probably a full-time job, whereas I've still got the same full-time job I always had, plus all the new stuff. I'm managing four farms, I'm doing the celebrity work, I'm doing the filming, and I've got a lot of charity work now. I've got about seven jobs now.

But by the time you've finished explaining that to them they're bored and want to walk off.

'Yes, yes, fascinating, darling. I have to go and get on this plane now, even though I have no idea who it belongs to or why it's here.'

It's a very odd feeling, meeting a lot of people who know who I am, when I don't know them. 'Hi Kaleb! How's it going? How's the farming? How's the missus?' 'Yeah, yeah, yeah, all good ...' But I still have no idea whether they're meant to be famous themselves, or a local who is vaguely familiar to me, or a complete stranger who's seen me on telly. But everyone's so nice, I don't really mind. I love the fact that I'm being approached by young kids, who give me a high-five and say, 'I'm going to be a farmer one day!' That's a wonderful feeling.

The only part I don't like is people who want to put their arm around you. I hate being touched. Really hate it. I'm surprised I've got a kid, to be honest.

'You're surprised?'

I've started doing a lot of charity stuff, because as soon as you become a celebrity, that's a massive thing. You've got this whole following, so you can be the face of a charity, which is amazing. I know I say 'amazing' all the time about this, but that's because it is. I give tractor lessons and do online auctions. I give a lot of talks to young farmers.

Public speaking is something I never would have thought about before, let alone actually done. Six years ago, I wouldn't have dreamed I'd be standing up in front of an audience and they'd want to listen to me. I was just a kid who wanted to be the most successful teenager in Chipping Norton – that was my level. There was no other world to me; it was just Chipping Norton and my egg round.

It's a bit like a paper round, only you should hear the comments you get when you chuck them onto the doorstep.

It's the most amazing feeling in the world when people come to see you and listen to how you got started. It's opened up a whole new world, a whole new role, where I'm trying to inspire young people and teach them about farming. And I'm not exactly ancient myself.

I just had a hard egg round.

I've taken naturally to public speaking, the way I did to filming. Jeremy says I'm the most natural person he's ever worked with on camera, and I know I give him some stick, but coming from him that's one hell of a compliment. With public speaking, it just flows on the night.

Do not, on any account, try that old trick of picturing the audience half-naked.

You've got to know your crowd. If I'm doing a young farmers' night I know for a damn fact I've got exactly five and a half minutes before one of them picks up their phone. Because they're young farmers and all they think about is driving tractors, and when they can't actually drive tractors, they'll start messaging each other about tractors. Which I get because I'm a young farmer too.

Do not, on any account, try that old trick of picturing the audience half-naked on a tractor.

You've got to hold the fort for five and a half minutes and then keep them involved by asking them questions, because young farmers love talking about what they do. Which again, I know, because just try shutting me up about my job.

I want to build up young farmers' self-confidence to the point where they can go to their boss and say, 'I think we should try it this way.'

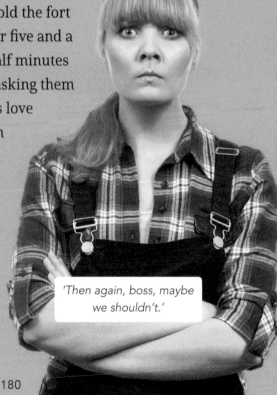

'Then again, boss, maybe we shouldn't.'

Before I became a celebrity myself, I thought a celebrity was just somebody who'd got a bit rich and full of themselves, and couldn't open an umbrella. They would have to get somebody else to do it, because they're incapable of everything. If I ever forget how to open an umbrella, then that's it, I've lost myself.

I hate it when people say, 'Don't change! Don't change!' I start thinking to myself, 'Am I changing?' Because I don't feel like I'm any different. Although I am older, so I suppose I must be. I'm twenty-three now, and everybody changes as time goes on. When people see me, they say, 'Oh, you've changed in this or that way.' I want to

'Don't worry, Kaleb, you can stand under mine.'

say, 'Well, that's what you do when you grow up. You never stop changing as you get older.' Am I supposed to stay exactly the same just because I'm a celebrity?

'I've been in this business for over forty years, baby, and I can tell you that's what the public want.'

The other day, there was a lady who – out of nowhere – showed me her tits. That never used to happen to me before I got famous, and frankly I'm not that keen on it happening now. To be honest, everything on her was very … south. But then everything on my end is very south as well, because, as I said, I'm maturing. Not that I'm about to show anybody. Unsolicited nudity is a celebrity perk I could do without.

After pulling her top back down, this woman said, 'I've just shown you my tits – can I have a selfie?' I said, 'Look, I'd have given you a selfie without seeing them.' Then her husband came over and asked for a selfie as well, and got her to take the picture.

I think I prefer it when people just freeze in shock when they see me. Or fall over, like someone did the other day, just from looking at me. I admit, I did laugh, but I also did go and help her up. I like helping people. That's really what I'm a celebrity for in the first place: helping Jeremy do all the things Jeremy doesn't know how to do. I'm not so much a celebrity as I am a celebrity babysitter.

'You can get rich and famous for this? When does that happen, please?'

'Hi, everybody, I'm your licensed Kaleb Cooper™ for the evening, I'm thrilled to be here!'

I'm not about to start doing personal appearances in nightclubs, either. I can't think of anything worse. If I ever get any celebrity lookalikes, I'll hire them myself, and let them do it as a franchise. I can take my cut of the money and stay at home, and they can get climbed all over by sweaty twelve-year-olds who got in using fake IDs.

KALEB'S TWO-POINT GUIDE TO BEING YOUR OWN BRAND

1. DON'T

People talk about celebrities being a brand, and I hate that. I'm like a walking billboard. When I go into a store nowadays, they don't let me buy whatever I'm after – they try to give it to me for free, then take loads of pictures of me holding it. Just take my money! I've done one advertisement, and that was for a fast-food chain, one I actually like, because British farmers do well out of them. That's why I agreed to do it.

It was when I'd first become famous, and I was trying to work out what it was all about, with nobody to help me. So I took a bit of money to do this advert. I said what they asked me to say, got paid, and off I went, thinking, 'Is this what celebrities do? That all seems a bit too easy.'

The advert went out, and it was the first time I'd ever experienced hate. 'You sold out! You're making farmers look bad! That company is bad for the environment!' And the advert wasn't even about burgers – it was all about biodiesel. I thought, 'Sh*t! I've done something massively wrong here.' Then, within two days, everyone had forgotten about it. And after a week, the advert got removed because it hadn't been properly fact-checked. So, it turned out to be a self-solving problem.

2. DO

If I have to be a walking billboard, I'm going to be my own billboard. Everything's a learning curve, and I've learned that advertising's not for me, unless I'm advertising my own brand. I've done cider. Cider's something I love – it's something all young farmers love – and it will really be my cider. I'll be picking the apples that go into it. Literally – I'll be picking them up off the ground in the orchard. So I'll be making a cider *and* tackling food waste. And I'm going into the factory to make sure it's all set up the way I want it to be.

The most important thing for me, though, is that by being a celebrity, I'm basically advertising what I love. I'm doing the two jobs that I love most in the world, TV and farming. And I love TV because I get to show to the world, to the lovely audience, what I love even more, which is farming. I want to inspire young people to go into agriculture, or to start their own business, or just to get up off the sofa and go and earn a few bob. Even if you have to start by scraping out muck.

Chapter Eleven

Partying

'Come on, lads – where's the cider?'

No.

Some people expect a party to involve guests standing around in their glad rags with a glass of wine in their hand, or sitting down to a fancy dinner. Well, that's not how me and my mates do it. Young farmers go hard, or go home. Strictly speaking, we do both, one after the other. Although the second one is optional, and often not accomplished on the same night. But if you can remember doing either, you probably didn't do it right.

Yes.

Also, the dress code is always the same. Women go to the party in what they wore for work. And so do men, except without washing the mud off first. There's no dressing up.

Every young farmer learns early on that there's nothing like a flat cap to wow the ladies.

You're expected to go in your work boots, your work jeans, a Schöffel – one of those fleece gilets – and a checked shirt. With maybe a flat cap, if you really want to ramp up the sophistication. It never changes. Which is handy, because you don't have to either.

You normally go to balls – young farmers love balls, and obviously there are a lot of T-shirts worn by young lady farmers pointing out this fact – which usually take place in a barn. You almost never go to a posh venue, which is why you almost never have to wear posh clothes. If you're throwing a party, you have it on your farm. 'We've got this barn clear, the grain's gone out – let's have a party before we start filling it up again.' That's how you end up having potato fights.

Wait, what?

'You should have seen the other feller.'

It happens because all the potatoes are stored in the barn before they go out to the supermarket the next morning. You get to about two o'clock in the morning, everybody's had a drink – by which I mean everybody has had all the drinks – one potato gets thrown, and that's it. Potatogeddon. The whole lot go everywhere.

So you stay in the barn, get a couple of hours' sleep, then collect the potatoes that have been strewn around, grade them and put them all back ready for loading at 7am. Half an hour's fun, and three hours' cleaning up. People don't realize their spuds were ammunition only a day ago. What they don't know won't hurt them. Although it bloody well hurt us, I can tell you. You can get quite a bruise off a well-aimed potato.

When people think of barn dances, they probably think of America, because they've seen it in the movies. But it happens all the time here. You've got the bales of hay or straw to sit on. Straw hanging from the pillars or the ceiling. Proper country music. Maybe even some line dancing. Although that's more the older generation of farmers now. With young farmers, we never get to that point – we're totally hammered long before that can happen. We're too drunk to do anything that organized.

In fact, there's never anything about a young farmers' party that's organized. I've been to parties where there were no lights at all. So, young farmers being young farmers, we think, what can we do? We get all the tractors in a massive circle with the headlights on, and there you go. People think all the acid house fans invented meeting up in a field for a boogie back in the eighties, but I'll bet you young farmers were doing it long before, and we're still doing it now. We're the true underground rave scene in this country.

Acid house? Acid barn, more like.

All our ones are probably illegal too, but no one else knows about them. They're in the middle of a massive field, or in a barn. The only people who might ever hear them or complain about them will already be at them. The party can go on as long as it wants. It doesn't have to stop at 2am. Sometimes fights happen. That's because two young farmers want to do things their way. Usually it's fights over who's got the best tractor. It's like a cock-fighting ring, only with farmers: John Deere goes up against New Holland, they fight it out, you place bets on who's going to win. You can make loads of money. It's been going on as long as there have been farmers. I bet they used to fight over who had the best scythe, or plough. And I'd find somebody to take that bet, too. I'm a bettor, not a fighter.

'Say that about my barrow again! Go on, say it!'

My plan is, as my contracting business grows – if that's not a contradiction in terms – I'm going to buy one of every different make of tractor. One Claas, one Case, one New Holland, one JCB, and so on. So when I go to these fights,

'I'll take the lot.'

whoever the winner is, I can always say, 'Yeah, I've got one of them – it's a wicked tractor.' It's all about improvisation, and staying on the right side of the argument.

When it comes to drinks, it's always cider, as you know full well by now. Pure cider. And cider is very, very, very gassy. So at a farmers' party, there is a lot of farting and burping. It's not

for the delicate. And there's always a lot of sick, because they just don't know when to stop. I went to a party last year called Moreton Mischief. To get into this party, you had to walk through a horse box …

Wait, what? Part II.

… and pay your £5. Horse boxes have two doors, one at the back and one at the side, so you can load forwards, and unload forwards too. So they just treated the young farmers like horses, which makes sense – we get easily spooked, and then we're liable to kick off. The site had Heras fencing, made of individual wire mesh panels, and about six toilets. The women refused to share the toilets with the men – 'No chance, they're dirty f***ers, they piss everywhere' – which was fair enough. So the men started pissing out through the fencing. As you approached, the first thing you saw was a line of willies poking through the panels, like maggots. Good thing there weren't any chickens around or it would have been a penis massacre. It's an image I wish I could forget.

And one the publishers do not intend to replicate. Here are our thoughts on the matter expressed through the medium of kittens.

It's not the worst thing I've seen at a party, though. That was when I was thirteen, going on fourteen. It was at a place called King Stone Farm, which is named after the ancient King Stone, part of the Rollright Stones. Witches and druids love those stones, and twice a year they meet there to dance naked.

'Look, do we tell you how to have fun?'

I was checking on the dry cows – the cows that aren't being milked, because they're going to give birth – around half past ten at night. I could see about twenty yards in front of me by moonlight. I was walking through the farm, carrying my little torch, when I heard the sound of clapping. Clap clap clap clap clap! I thought, 'What the hell is that? Is a cow stuck, or something?' So I walked to where the noise was coming from, shone my torch up ahead, and lit up the sight of a guy mounting a woman while a load of druids applauded.

It was my first encounter of sex – in humans, that is – and I'm surprised it didn't put me off for life. Say what you like about young farmers, but at least we don't have parties like that. Not until it gets after eleven, at any rate.

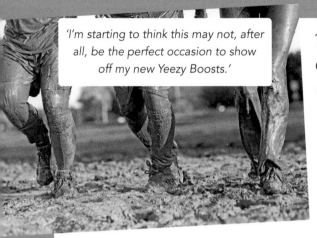

'I'm starting to think this may not, after all, be the perfect occasion to show off my new Yeezy Boosts.'

Those who call themselves druids and witches aren't always from the countryside, so in a way, it's just more townies making a show of themselves. And you'll know from what I told you in Chapter One that you can easily pick out the city people at any farmers' party. You can see them coming a mile off, trying to pick their way around the puddles, in their white trainers and white shirts and nice jeans.

They stand out even more than druids would. They are not dressed to be rolling around in a pigsty, fighting the pig. Yeah, that stuff does happen. I wonder who invented that particular game? It definitely wasn't the pig – but at least the pig is used to wallowing in mud. It's f***ing stupid. But we all know that we're all going to have a go at it. So, if you don't fancy that, you're at the wrong party.

'Come and have a go if you think you're hard enough.'

It's fine that people want to experience a young farmers' party. It's just that they don't always fit in with the whole situation. They don't know how to act. They find themselves thinking, 'Oh. These are quite ... broad people who want to get their opinion out.' And young farmers can come across like that. But they're so soft. They don't ever mean any harm to anyone. Still, you don't want to pick a fight with a young farmer, unless you're another young farmer. They're built like a brick sh*thouse. They work hard for a living on the land, lifting milk buckets and

Ditto you, if you start anything.

chucking sacks of grain around. They're massive. And they're all down for a good time because they don't get off the farm very often, so they're making the most of it to get absolutely drunk off their face. They work 24/7, and the one time they get to party, they're going to hit it hard.

Also, young farmers always treat everyone the same. At a young farmers' party, nobody cares how much money anyone's got – you're all equal, and all there to have a damn good time. Even if they haven't got any money themselves, they're not tight. Young farmers will always buy you a drink.

As soon as a party's planned, word travels as fast as a pig can get in his bowl of food. Although that pig may well end up on a spit. There's always top catering.

'I've had the time of my li-i-ife …!'

Then, after a few drinks, you get an idea in your head, like getting out the quad bikes and dragging around an old tractor bonnet for people to ride on. It hurts like a bag of sh*t when you fall off. I cut my knee open. My brother landed on his face. And we agreed it was the best fun we'd had all year. There ain't no party like a young farmers' party. And that's probably best for the preservation of the human race, because I'm not sure anybody else would survive it.

KALEB'S GUIDE TO HISTORY'S MOST NOTORIOUS PARTIES

EDWARD RUSSELL'S PUNCH PARTY (1694)

This guy invited 5,000 sailors to a party, which sounds very liberated for its time, until you find out he was an admiral in the Royal Navy. They had a fountain full of brandy-wine punch served by a kid in a canoe paddling around in it. I mean, I know farmers aren't always the best when it comes to observing health and safety regulations, but I'm not sure that even we would go that far.

'Next, I propose we fill the English Channel – here – with Pimm's.'

ANDREW JACKSON'S GREAT CHEESE LEVEE (1837)

I didn't know there was a US president madder than the last one, but old Andy here is in with a shout. A local farmer presented him with a 1,400-pound wheel of cheese – proving, once again, how farmers are the unsung heroes of history – so he threw a huge party and invited anyone who fancied it to eat it. I would have been there like a shot. I'm into that any day. Although I do worry about gout. A fair few older farmers get it, and the amount of meat and cheese I eat, I might be in line for it one day. But no way would that have stopped me. This is one party I would one hundred per cent not miss.

'Now all we need is a 5,000-litre wine box.'

'Neigh, neigh, and thrice neigh!'

BIANCA JAGGER'S THIRTIETH BIRTHDAY PARTY (1977)

I thought Bianca Jagger was married to one of the Rollright Stones, which was confusing but not all that surprising, considering what goes on there. But it turns out it was the bloke in that 'Moves Like Jagger' song. Anyway, she rode a horse around a disco, and apparently this was a big deal. But we do that sort of thing at our parties all the time, only usually with something more dangerous – see how long you can stay on a cow or a pig before you get thrown off and break something.

One girl I know, who's on the small side, rode an Irish wolfhound twice around a pub car park. Let's see Bianca do that before we start getting all impressed.

Chapter Twelve

Travel

Onwards I venture, into the dark heart of Oxfordshire.

Kaleb's horizons in the old days.

Travel broadens your mind and expands your horizons. Before I was on the telly, I'd only once in my life been out of Chipping Norton, and that was the school trip to London when I stayed on the coach. But now that I'm a celebrity, I've turned into a jet-setting globetrotter. Banbury. Bicester. Tewkesbury. I've seen them all. I even took a trip to the bright lights and the fleshpots of Oxford, although that was all a bit too much to take in. I'm boldly going where no Cooper has ever gone before. I don't mind going further afield, as long as wherever I go has an actual field in it.

Kaleb's horizons now.

Buses, trains, planes, ferries, ships, taxis – I've never been on any of them. I wouldn't know how to catch a bus. Do you have to throw yourself in front of it, or something?

Only at a request stop.

I've started paying much more attention to travel now. When I drive from Chipping Norton to Chadlington, it's 3.7 miles. It takes me six minutes. Six and a half on a bad day, if school's on. I go past two farms and eight fields. The rotation in the fields is wheat, barley and rape.

That's the kind of journey I'm comfortable with. I don't want to start doing mega-voyages all of a sudden. I'd rather go a little bit further each time. Maybe two miles a week. I reckon at that rate, it'll take me about eight months to get out of the country. Or at least, into the sea, although what I'm supposed to do there, I have no idea.

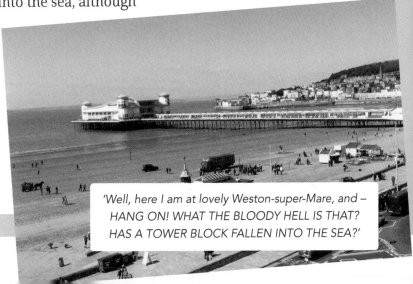

'Well, here I am at lovely Weston-super-Mare, and – HANG ON! WHAT THE BLOODY HELL IS THAT? HAS A TOWER BLOCK FALLEN INTO THE SEA?'

My problem is that I want to see the world, but I don't ever want to leave the farm. So I've worked out a solution: I'll go and look at loads of other farms all over the world. I'd love to go and see the cattle farms in America. They don't do cows in hundreds, they do them in the tens of thousands. And the ranches can be over a million acres. It reminds me of one of my favourite farming jokes. A Texan farmer is boasting to an English one: 'I've got a truck on my place, and I can get into it in the morning, drive all day, and by sundown I still haven't reached the end of my land.' The English farmer says, 'Yeah, I've got a truck just like that, too.'

'Snap!'

In South America they still have proper cowboys, who apparently are called grouchos, which seems a bit weird, but there you go. Then again, with the price of red diesel being what it is now, I'm thinking, maybe it's time to do everything the old-school way again. Not just beef and dairy – we're going to get in the spring barley by horse.

'Olé!'

'Nah, mate, this is the small field. The big field's over there, in Queensland.'

Everything's just so much bigger in other countries. Like grass cutting in New Zealand. I want to go silaging there. And harvesting in Australia – the size of it! It's massive. I can picture them watching our show and thinking, 'Look at them, with their one tiny little combine harvester, taking eight hours to cut their tiny little field. Isn't it adorable?' And they've got thirty combines in one field, and the field is twenty-five miles long.

Another thing I've heard they do in Australia is herding livestock by helicopter, just because the place is so big. That's even cooler than being a cowboy. Let's say you've got 370,000 acres and 20,000 cattle, like one guy I've heard about. You can do in a week what used to take more than a month. I definitely fancy trying my hand at that, as long as somebody else pilots the helicopter.

Not you.

Even more bonkers than that is being a nomadic herder in Africa, where the grass follows the rain, the cows follow the grass, and you follow the cows. It's a bit like the way they used to farm sheep in Yorkshire, with the shepherds following them around and staying in little wood and stone buildings for a couple of nights at a time. Only in Africa, you have to go for hundreds and hundreds of miles, and the place is full of things that want to eat you.

I'd draw the line at that one. Imagine having your five lovely cows and needing to protect them from a f***ing lion. I mean, I have to protect my rape crop from pigeons, and even that tests me to the limits. I'll fight a pigeon any day if I have to, but I wouldn't fight a f***ing lion.

Potayto …

… potahto.

211

Something I'd love to go and do that's totally different from what I'm used to is rice farming in South Asia. That's hardly changed in centuries. It's like the opposite of grain farming. Grain farming you can do with not that many people, but a lot of machines. Rice farming is all about manual labour. There's not enough money in it to buy the machines, unless you go really big – although they do have rice tractors with amazing spiky wheels for the people who can afford them.

I admit, if you ask me my idea of a good time, farming in two feet of water isn't the first thing I'd think of. But I'd love to give it a go anyway, just to find out. I really like how, instead of chemicals, they use ducks to get rid of the weeds and pests. Then, when the rice harvest is in, you can eat the ducks with it. A perfect combination.

'Hang on, you can do what?'

I want to go to India, too, but not to do any actual farming. They do a lot of shifting cultivation there, where you farm the land to exhaustion, then move on somewhere else and leave it to recover on its own, which isn't great. I get that it's what some farmers do when they're trying to scrape together a living, literally, but I wouldn't want to be a part of that.

What I do want is to see the Rural Olympics. There's a place called Qila Raipur in Punjab which I'm sure would mean a lot more to me if I knew where Punjab was, but I do know one thing about it, and that's that the people who live there must be the hardest, and most insane, farmers on the planet. And when you look at farmers generally, that's a pretty high bar.

It's not just that they do bullock-cart racing that makes the chariot scene from *Ben Hur* look like a ride on the dodgems. It's not just that they pull heavy machinery around with their hair – what the f***?! Nobody's trying that with my hair, I can tell you. It's not just that they have every kind of tractor-related event you could ever conceive of. Tractor tyre-rolling – which you'd have to be nuts to do for fun, it's incredibly hard work – and tractor races, which I'd go for every time.

It's that they have a contest where people voluntarily get run over by tractors. It's basically Extreme Farming. It's as if they've looked at farming – which, you know, is difficult and dangerous enough already – and gone, 'How can we make it more difficult and dangerous? Wait, I've got it!'

'Call that a tractor, you lightweights? C'mon, why don't you run me over with a proper machine, not some child's toy?'

In the Mediterranean, they get wet winters and dry summers, unlike here, where we get wet everything, all the time, until the actual moment you need it to rain, as a farmer, at which point it stops. They grow a lot of citrus, olives and grapes there. The trouble is, I don't like any of those things. I don't like wine, either – if it isn't cider, I'm not interested. So there's not a lot for me there.

But the one thing I do like is the way they harvest the grapes. Everybody thinks it's a bunch of happy, singing people who look like models plucking them off the vines by hand, but it isn't. It's a thing that looks like a war machine from a sci-fi movie that goes either side of a row of vines and shakes off the grapes into a pan. I totally want to get a closer look at that, as long as they don't try to make me eat any.

Grape harvester, or Graxxan mutoid cyber-tank from Irregulon Six?

This is all very ambitious for somebody who's never travelled all that much. And by 'all that much', I mean, if I couldn't ride there in ten minutes on a tractor, then it wasn't ever going to happen. Perhaps I should start a bit closer to home and go to London again to see the underground farm they've started there. It isn't full of hipsters with funny-coloured angular haircuts who'll lose interest as soon as the farm goes mainstream. It's actually underground, in a huge bomb shelter left over from the Second World War, growing fruit and veg with hydroponics and LED lighting. Which is way, way more cool.

Not the platform at Oxford Circus.

Speaking of which, there's also a greenhouse in Antarctica that can grow food when it's -70 degrees outside. I would just like to say, here and now, that I am never, ever going to visit that. Bits of me would break off. Bits that I like and use and intend to hang on to.

KALEB'S GUIDE TO THE FARMS THEY CAN'T PAY HIM ENOUGH MONEY TO TRAVEL TO, ALTHOUGH THEY'RE WELCOME TO TRY

WELSH LEECH FARM

This is even worse than sheep. Say what you want about sheep, but at least they don't suck your blood. And if they ever do invent vampire sheep, that's the day I start campaigning against genetic engineering. I'm sure there are good medical reasons for farming leeches, but it's not a risk I'd ever be willing to take. 'Just going to feed the leeches, love.' And a minute later you stumble back in looking pale. OK, in my case, paler. 'They got me, darling! I've lost eight pints of blood, and I only had ten to start with.' What farmers will do to make a living.

Nope nope nope nope nope.

CANADIAN MARIJUANA FARM

We grow hemp over here, which is the same thing, only it isn't strong enough to smoke. That's fine by me, as I've got zero interest in any drug that isn't cider. We have to get a licence to grow it, it can't be near a road, and we've got to have a contract to sell it straight away. But it looks like a field of weed, because that's what it is. Meanwhile, in Canada they're growing vast amounts of it in remote areas – which is pretty much the whole country – and it's strong enough to knock out a herd of bison. It's crazy how you can make £1,800 a tonne from it. But it's not going to feed the world, is it? Just the opposite. It's going to give the world the munchies.

That explains why they can't stop listening to Rush.

'Reckon we're more into cage fighting, mate.'

AUSTRALIAN KANGAROO FARM

Another one that's got to be worse than sheep. At the moment, they don't farm kangaroos, they just shoot wild ones. But they're about to start, and it's a no from me. Sheep can't bounce and they can't box. You go out to bring the kangaroos in, and you come back fifteen minutes later looking like you've gone five rounds with Mike Tyson.

AFRICAN CROCODILE FARM

There's an urban myth – a rural myth, really – about a farmer who goes out to feed his pigs, collapses from a heart attack, and the only thing they find left of him is his wedding ring. But with crocodiles, you wouldn't even find the ring. There's a reason why they've been around for millions of years and outlived the dinosaurs. Do not f*** with them. Crocodile farms? F***. That. I couldn't think of anything worse. I must confess, though, I would like to see one. But from a distance. I'm thinking around five miles should do it.

You might want to move back a bit. Then again, you might want to check what's behind you first.

'I can see your farm from here!'

THE INTERNATIONAL SPACE STATION VEGGIE FARM

You've got to be impressed by anyone growing salad leaves in space. That is pretty cool. But there's farming in lovely places, and there's farming in space, and I know which one I'd rather be doing. I'd rather be in the middle of the countryside in Chipping Norton. I want to look out my tractor window and see rolling hills with deer running through them. I don't want to look out the tractor window and see the world spinning in a void of darkness. In the end, there's no place like home, and I'd rather be in it than looking at it from hundreds of miles away.

ACKNOWLEDGEMENTS

First, I want to thank Jeremy Clarkson, without whom none of this would ever have been possible. Thanks to Peter Fincham, Andy Wilman, Zoe Brewer, Alice Gordon-Lyons, Conor Tighe, Peter Richardson, Gavin Whitehead and the entire *Clarkson's Farm* team at Expectation; to Kate Killgour and Lou Plank at Plank PR; to Dan Grabiner, Sophie Spirit and Charlie Martin at Prime Video; to Debbie Catchpole and all the team at Fresh Partners; and to Jane Sturrock, Katy Follain, Andrew Smith, Charlotte Fry and Georgina Difford at Quercus Books. Thanks also to Julyan Bayes for designing the book, to Elaine Willis for her tireless picture research, and to David Bennun for translating the contents from the original Kaleb into Fancy Literature. If there's anyone I've missed – sorry! There's so much going on that it's been hard to keep up, but I'm grateful to every single person who's helped me out and made this book happen – along with everything else.

Picture Credits